Eden 2.0

Mitchie Gonzales

Eden 2.0

ISBN

Paperback: 978-621-10-2144-7
Ebook: 978-621-10-2142-3

Book layout by **JK Ojano**
Book cover by **Cyrill Bagos**

Published by 8Letters Bookstore & Publishing
Philippine Tel no: 0915 268 8718
International Tel no: +852 6690 5941
Website: www.8lettersbooks.com
Company Address: Marigondon, Lapu-Lapu City, Cebu. 6015

To Jad for never asking why I have to write

*To Maya, may you reach your dreams,
stay free and soar to new worlds*

*And to those who believed before,
may you still believe now*

Contents

Prologue

When the woman saw that the fruit of the tree was good for food and pleasing to the eye, and also desirable for gaining wisdom, she took some and ate it. She also gave some to her husband, who was with her, and he ate it.

—Genesis 3:6

After-Earth (AE) 722

"AHHHHHHHH!" the blonde woman screamed in agony. She was naked in the warm tub, with the water reaching her breasts. It was stained with blood as she lifted her back, screaming as she did so. She was nearing the end of her labor. She shuddered in excruciating pain, which was almost blinding. She tried to purse her lips, as her breathing came out in short, shallow bursts.

"Push," a feminine robotic voice instructed simply. It came from an elderly woman hunkered down beside the tub. She had gray hair tucked in a bun and wore a white uniform. Her movements were stiff, and her eyes were focused on the girl. Her face held not much emotion.

Watching them, not too far away, was a man pacing about the room. "Is she going to be okay?" he asked the woman. "Can we have a real doctor?"

As if on cue, the blonde woman screamed again, pushing hard as she did so. Her face contorted in pain, and her eyes were squeezed shut. Her knuckles turned white as she grasped the edges of the cold marble tub. The water slushed around the edges, spilling onto the cold marble floor.

The old woman's head snapped to the man. "I am the doctor. My body is AI, yes, but rest assured, I am on top of things. Your wife is doing right in the water birth."

The man understood that the AI doctor was angry, but her robotic tone almost did not give away any emotion. Almost. The man sighed, irritated.

The loudest and the longest scream finally escaped the woman's lips, and it eventually gave way to several episodic gasps.

The AI doctor reached down. The man peered and gasped. His child, finally. "Oh, oh!" He held out his arms.

"A boy," the old woman announced. Tiny cries erupted as she scooped the infant out of the bathwater.

The blonde woman let out another painful scream, but this time, it was shorter. The old woman reached down again. She stared. "Oh, heterochromia," she simply said. She cradled another infant out of the bathwater and presented it to the man.

"A girl!" the man gasped, peering down on the baby.

"She has to be given to the intensive care nursery," the old woman mechanically announced.

"What? Why? Why?!" the blonde woman began to ask hysterically, her wet arms trying to grab her newborn.

"What's wrong with our girl?" the man demanded.

"Heterochromia," the old woman said robotically. "Her eyes."

The couple peered. The woman reached for her baby. The left eye was green, and the other was light purple with a blue tint.

"Complete heterochromia," the old woman repeated. "A genetic mutation. Mutations are prohibited in Eden 2.0."

"No, no," the man tried to protest. "Some mutations are harmless."

"Mutations are prohibited in Eden 2.0," the old woman repeated robotically, looking at the man. "This anomaly will be sent to Intensive Neonatal Anomaly Care. This anomaly will be treated, and you will be contacted when you can get it. Then you can name it and nurture it like a child."

"It is a child!" the blonde woman angrily insisted. "It's a *she!*"

"Mutations are prohibited in Eden 2.0," the old woman repeated for the third time. "Protocol must be followed." She stood up and straightened her back. She held out her arms for the baby.

The man and the woman exchanged a look. They knew what must be done.

Chapter 1

Eighteenth Birthday

The Lord God planted a garden towards the east, in Eden; and there He placed the man whom He had formed.

—Genesis 2:8

Violet Malaya rang the doorbell of a white glass house. The dark glass door lightened and opened instantly.

"Violet," a warm woman with wavy brown hair cascading to her shoulders greeted her. "The twins are just getting ready. They will be out in a few minutes."

"Mrs. July." Violet nodded. "It's okay, I can wait." She twirled her black wavy hair with purple highlights. She wore her usual black suit, but her bathing suit was underneath.

It was the twins' eighteenth birthday. In Eden 2.0, it was a tradition to swim in the Sophronia Ocean, the biggest body of water, on your eighteenth birthday.

They heard footsteps, and both looked to see Mr. July, a tall, handsome blond man in a business suit, flanked by two teenagers.

Alec and Exa July were fraternal twins. Like their father, they were blond. Exa's hair was curly and fell onto her shoulders. They had dark blue eyes. Despite being opposite sexes, they looked so alike. They smiled the same, talked in the same manner, and had the same mannerisms. They were both quiet and kept to themselves.

Violet had been friends with the twins since kindergarten—in fact, it had only been them since then. They were on good terms with other teens, but the twins just preferred the company of a few people.

"Violet!" Exa greeted.

"How's eighteen?" Violet smiled as the three figures reached the door.

"Never been better," Alec smiled at her as he swung a duffel bag on his back.

"I really don't know how this tradition came to be," Mr. July said. "But I hope you kids have fun on the beach. Not too far—"

"The beach?" Mrs. July snorted. "This is the dive. In the *Voguer W*. All eighteen-year-olds do it, dear. They'll be in the middle of the ocean."

"That's what scares me!" Mr. July shuddered. "The deep ocean. You kids paid attention to school, right? They taught you how to dive?"

The three teenagers nodded. "Relax, we got all we need here," Exa said reassuringly, patting Alec's duffel bag.

"You sure, son?" Mr. July turned to Alec. "Everything in there?"

Alec nodded. "Yep, checked them twice!"

Mrs. July chuckled. "Everything will be fine, Harold. Stop fussing!"

Mr. July pretended to sniff. "Can't believe they have grown up so fast!"

Mrs. July placed an arm around her husband's waist and smiled at the three teenagers. "Go and enjoy your day," she said.

"Just call us if you need us!" Mr. July added before hugging the twins.

The twins stepped out of the house.

"Game?" Violet asked. The twins nodded.

The three looked at their wristbands. It was mandatory for all residents of Eden to have one. Everyone got one by the time they reached seven years old. The wristband was their means of transport, their wallets and banks, their phones, their social media, their news, and everything else. They pressed on their bands. Three hologram screens flashed up from the bands.

"The ocean or the island?" Exa asked to confirm.

"The ocean," Mr. and Mrs. July suddenly interjected in unison.

"If you're thinking of going to Templos," Mrs. July started.

"Then forget about the whole thing!" Mr. July finished.

"Those kids up there are up to no good! Do you know you can break your spinal column jumping at that height?"

"We have teleportation," Alec reminded their dad.

"Yeah, it's like a parachute." Violet shrugged. "Teleport before your head hits the water's surface."

"No!" Mr. and Mrs. July said sternly.

Violet looked down. The twins sighed.

"All right, we'll be good. We'll head for the ship," Alec

to meet him?" He winked at Violet, who smirked.

"A schoolmate?" Exa asked. "We know almost everyone."

"Oh, he's from Isle Luna." Rufus looked at them wide-eyed.

Alec raised an eyebrow. "So?"

"So." Rufus turned to him, acknowledging him for the first time. "He's a goddamn foreigner. We got to make him feel welcome!"

Exa looked up. Sophronia Ocean spread out everywhere, and their ship was in the middle of it. Upward, she spotted the floating islands from a distance. Artificial intelligence made the floating islands possible, like almost everything in Eden. Technology created a new paradise. AI could make floating islands with floating waterfalls by tweaking the gravity at specific atmospheric points. Just think of everything arranged in a grid, like calculated plots of time and space. Everything was AI.

In fact, the ship they were on didn't really travel because AI had given everyone teleportation.

How many islands were in Eden? A lot. But despite that, there were no nations in Eden. There were no passports. If one knew of a place and the wristband allowed it, one was more than free to go there.

Money was almost nothing to worry about. They had banks, yes, but wealth in Eden was abundant—it was practically nothing. Just look at this ship. The *Voguer W* was an everyday party ship for every eighteen-year-old.

There was no race in Eden. The moment a woman was expecting, AI doctors asked the parents what skin color, hair color, eye color, or other traits they wanted their child to have.

Eden held your genetic identity. It could change anything, and when you were born, you became the change Eden had created, its ultimate design. But what comes out different than expected is an anomaly, which is strictly prohibited in Eden.

Exa stiffened. Rufus tried to put his arms around her and Violet. Alec's eyes narrowed.

"You know Isle Luna, right?" Rufus said to Violet. "You're the top one at school, Violet, right?"

"They're into witchcraft." Violet shrugged Rufus's arm off her. "But everything can be explained by the AI era." She rolled her eyes. "Besides, what's so special? We've met a lot of people from Isle Luna before."

Religion was one of the very few things Eden got from Earth. There were lots of religions in Eden, including Christianity, Islam, and, well, witchcraft. However, it was one of the most common triggers for debate in Eden, and several groups had come up to abolish the idea of religion: the atheists. Still, the majority wanted to fight for their faith.

Rufus frowned and shrugged like some loony. "Oh, okay, I thought you'd be interested in knowing that he has some big connections and knows many secret places in Isle Luna. He also just turned eighteen recently too."

Even Exa, who was calmly enjoying the view, smirked. "That should interest us because?" Her and her brother's wristbands had added several places they could visit, like Isle Jungles.

Eden knew everyone's birthdays. Their wristbands had always greeted them promptly on every birthday. When they turned seven, they were gifted the ability to teleport into many isles with different play areas and museums. Today, Exa and Alec were given almost fifty more isles to explore.

Rufus smiled and whispered to the three, "He can fly."

The three teenagers looked skeptically at Rufus. Even Alec looked interested. "How so? Teleportation is the only ability Eden gives to its residents."

Rufus smirked at him. "If you're curious, we will be at Isle Templos. He'll show the other cool kids there how to dive by flying."

"You mean jumping." Alec sighed. And he thought this was interesting.

Rufus clapped his back. "Alec, it's flying, I swear. He takes off his wristband—"

"WHAT?" Violet exclaimed. "That's illegal!" she whispered loudly.

Rufus shrugged. "He never got caught—"

"Eden knows everything," Exa added in a stern tone.

Rufus shrugged again. "I don't know, okay. He takes off his wristband. He cannot teleport midair after he jumps, and then he flies. Not a wet hair on his head."

"I thought he just turned eighteen," Violet said skeptically.

"He's been doing it since he was seventeen." Rufus smiled, knowing he had gotten their attention. He smirked at the three teenagers. Violet, a know-it-all, and these twins, who think they were mysterious, well, they thought they were so above other kids, huh. It was hard to get them interested in anything.

"If you're curious, you can join us," he invited them,

smiling at the two girls. He just really wanted Exa or Violet. The more girls tonight, the better the party!

Then he looked at Alec. "But if you're scared—"

Alec, who had seemed irritated since he had gotten onto the ship, said, "We'll be there, Rufus. But we will only watch this boy of yours fly."

Rufus scoffed. "See you, twins!" He pressed his wristband. A hologram appeared.

ZAP! Then he vanished into thin air.

Exa and Violet raised their eyebrows at Alec.

"I thought no Templos," Violet teased him with a smile. "We promised your parents."

Alec let out a sigh and a tiny smile. "I'm curious," he admitted.

"You just wanted to try the cliff-diving and teleporting thing," Exa accused, elbowing him. Alec let out a laugh.

"I guess we're going, then." Violet shrugged. She looked around the ship. There were groups of dripping wet teenagers laughing, gossiping, and raising their glasses, singing happy birthday. Some have started playing drinking games. "Hmm, maybe the party is more interesting up there."

The three pressed their wristbands. *ZAP!*

Chapter 2

Isle of Templos

Now a river flowed out of Eden to water the garden; and from there it divided and became four rivers. The name of the first is Pishon; it flows around the whole land of Havilah, where there is gold.

—Genesis 2:11–14

ZAP!

Alec, Exa, and Violet found themselves on top of a cliff, overlooking the vastness of the Sophronia Ocean. The three friends held their breath.

"It's beautiful!" Exa gasped.

"It is," Alec quietly agreed.

Violet beamed at the twins.

The moon in Eden was pink and so big that one could even see its craters. One could see more than half of it in the horizon of Sophronia Ocean.

And upward, at its height, was the sun. The sun in Eden was warm but never scorching. Like everything in Eden, even celestial bodies were controlled by AI.

The presence of the sun, almost side by side with the moon, in broad daylight should have been contrasting. But together, now, they looked very complementary, making the horizon a mix of bright light pink amid bright blue skies.

Breathtaking. Violet thought there was no other word for it.

Just a few meters in front of them was the river of Templos, turning into a waterfall as it ran over the cliff.

"I see you found the place," an arrogant familiar voice said. Rufus was walking toward them. He was with other boys and two other girls who looked like trouble. They were sneering at the newcomers.

"Thought these goodies didn't have Templos on their wristband," another guy said, smirking.

The Isle of Templos was one of the many uninhabited islands of Eden. It wasn't a big island. But what made it so interesting was that it was mysteriously abandoned a long time ago. No one knew why. A large ruined temple stood in the center of the island. It was declared off-limits due to dangerous infrastructure. But from the temple rose a natural well of fresh water. No one knows where it came from, but it eventually became a river. The river cascaded into falls, going down into the air, and creating a mist, tiny speckles of water joining the Sophronia Ocean. Its shroud of mystery and danger brought mostly teenagers to the island, cliff-diving and transporting themselves midair.

"Everyone is welcome on Templos," someone behind the teenagers spoke. He had dark skin, and his hair was arranged in long dark locks. His gaze was intense as it met Exa's. Exa smiled slightly. It didn't help that this guy was shirtless and rocking six-pack abs.

"Hey, Theos!" Rufus greeted. "I just told them about you!" He turned to Alec, Exa, and Violet. "This is Theos from Isle Luna."

Theos didn't look at Rufus. Instead, he continued to walk forward until he was in front of Exa. "And who might you be?"

Exa blushed. "Exa." She smiled up at Theos. He was tall, almost two feet taller than her. From the corner of her eye, she saw her twin roll his eyes. "And this is my brother, Alec."

"Twins," Theos noted. He reached to Alec with a fist pump. "Yo, man," he greeted. Alec met the fist pump without much of an expression.

"And this is Violet." Exa pulled her friend beside them. "My brother and I turned eighteen today."

Violet almost smirked at Exa. It was so obvious who was suddenly developing a little crush. She looked at Alec, who was looking around the place. She noted the two other girls– a red-haired one and a brunette one—eyeing Alec. She frowned.

"That's great!" Theos clapped Exa's and Alec's shoulders, surprising Alec a little bit. "No better initiation to adulthood other than…well…adventure."

Alec's eyes narrowed. "Rufus said you could fly. I'm skeptical."

Everyone laughed, including Theos.

"This one wanted to see action," the red-haired girl almost purred. She looked older than eighteen.

"There's also another better initiation to adulthood," the brunette one said, and they giggled.

Alec playfully smiled at them. Looking at these ladies, Exa knew they were too superficial for Alec. Alec was…deep. Way deeper than her. He was a poet and a musician. But both of them craved adrenaline. They were into sports in school as well. But again, no one really got close to them except for Violet.

"What would you like to know?" Theos turned to Alec.

"How do you fly?" Alec asked.

Theos shrugged. "I've been doing it since seventeen. And I turned eighteen last week."

"So how?" Violet repeated the question.

Theos bit his lip. "Some say Luna, the moon goddess, granted me the ability—"

"No," Violet scoffed. "We're in the age of AI, Theos. Whatever your religion can do—"

"Paganism and witchcraft are way above AI," Theos tried to explain, but the three teenagers raised an eyebrow.

Theos sighed. "I'll prove it to you." He suddenly took off his wristband.

Violet gasped loudly. The twins looked in shock. The other teenagers were surprised, but they started to smile and clap.

"Show AI who's boss!" one shouted.

Theos winked at Exa, who blushed, and then marched off toward the river. There were several hoots of encouragement and excited murmurs.

Theos, knee-deep in the water, trod until the edge of the cliff. He was holding on to the rocks he passed by. The water didn't flow too fast but wasn't steady either. There was moss, so the rocky ground might be slippery.

All the others, including the twins and Violet, followed Theos. Some went into the river as well.

When Theos reached the very edge of the island, he looked back at the twins and shrugged. His right hand pushed the last rock he was clinging to, and he pushed his body forward into the air.

Exa and Violet gasped. Alec's eyes widened in

anticipation. He walked quickly to the edge as far as he could go without falling, and the others did as well.

Down the cliff—or the island—Theos was almost a tiny speck of dust amid the misty clouds and the dark blue watery depths. Then there he was! Arms outstretched, Theos went up instead of down, gliding the clouds.

Alec couldn't believe his eyes. He thought he would be watching a guy fall to his death. "How?" he whispered.

"There has to be an explanation," Violet muttered beside him.

The others clapped and cheered, including Exa.

Theos flew upward, back to the edge of the cliff, almost gracefully. He had no wings, no parachute, and no any other mechanism.

"How?" Exa said breathlessly. "You were wonderful!"

Alec rolled his eyes at his twin. *Please, don't sound so smitten*, he thought.

Theos's feet touched the bare ground. Those in the river went out to meet him. They began clapping his back and shaking his hand.

"It's a miracle!" they were saying. "AI has been beaten!"

Theos's eyes met the twins' and Violet's. "A gift from the goddess of the moon," he said, smiling at them.

Violet started shaking her head wordlessly. It couldn't be. AI had never been beaten. It was an intricate technology. It was the present and the future. It was calculated. Yes, it had loopholes, but they only paved the way for more advancements. What was this trick?

Alec couldn't wrap his head around the notion either. He

looked at Violet, who met his eyes. But it seemed like they wouldn't be getting a logical answer now.

"Perhaps someday you can go with me to Isle Luna. I'll show you how wonderful it is," Theos continued. "Maybe you'll understand it more."

The others hooted.

"Yeah!" Rufus clapped. "A trip to Isle Luna! Who's up?"

The teenagers started to clap and hoot again. Even Alec started nodding. It might be worth a trip. Exa was smiling. Only Violet looked wary.

Theos walked over to Exa and held her hand. Alec rolled his eyes again. Violet wanted to laugh at him but bit her tongue.

"But for you, happy birthday." Theos smiled. "I'll guide you in braving those depths and transporting just above the surface."

Exa giggled. "Sure. Show me." She raised her wristband.

Violet found herself aggressively pushed aside by the other two teenage girls who immediately approached Alec. "I can show you," the red-haired girl giggled at him. "By the way, I'm Freia, and this is Sandra."

Violet frowned as she watched Alec smile sweetly at them. "Sure! Did you just turn eighteen too?"

The girls giggled. "Yeah, awhile back." Sandra smiled and touched Alec's shoulder.

They raised their wristbands.

"You don't need to wait until you're eighteen to try," Rufus suddenly whispered. Violet jumped. She didn't even notice Rufus beside her.

Other teenagers had gotten beers out and were holding

cups. Someone had turned on the music.

Violet looked back at the twins.

"They're having fun." Rufus shrugged. "Shouldn't you too?"

Violet sighed. "All right. Show me."

The two raised their wristbands. "It's simple," Rufus said. "Turn on your altimeter." His voice suddenly turned serious.

Moments later, the three teenagers found themselves on the edge, knee-deep in the river, gazing below the falls, the strands of clouds and the dark blue waters beckoning them.

Alec had dropped his bag nearby. He was naked, wearing only his shorts. Violet and Exa were wearing swimsuits as well. Their wristbands were ready and set to the altitude just above the water's surface, where they would teleport midair.

At first, Violet raised an eyebrow at Rufus, who explained it to her. Places, such as islands or buildings, that allowed direct teleportation were identified by the wristband for good measure. Eden provided them the safest entry point to that place. Putting in blindly random coordinates, much more altitude points, was as dangerous as teleporting oneself in front of a bullet, for example. It was risky. Eden discouraged it.

This is how men die in Eden, Violet thought. From recklessness.

Suddenly, Theos stepped beside the twins and grinned. "I'll fly into the depths with you."

Exa smiled back. Alec and Violet ignored him.

"One! Two! Three! Go!"

The four teenagers lunged into the air, spreading their arms like birds.

"AHHHHHHH!" They found themselves against the wind, but gravity was way stronger.

The wind was cold on their faces, and the wisps of clouds slapped at their cheeks with their mists. It was exhilarating!

Exa thought her heart was dropping to her stomach, but the thrill filled her up.

"WOHOOO!" Alec shouted in excitement.

Only Violet was silent, staring into the depths where she would fall.

Then Exa realized the ocean was getting nearer, too near. They were going too fast!

"NOW!" Violet screamed.

The three immediately pressed their wristbands to their preset altitude.

ZAP! ZAP! ZAP!

As planned, they were gone midair, reappearing just a second before their heads hit the waves.

SPLASH!

At her excitement, Exa forgot to close her eyes and saw the exact moment her head broke the ocean's surface. She saw the bubbles clouding her face, her vision. Then, the depths were suddenly too dark. The water hit her nostrils, stinging her. She forced her way up, coughing when she did so.

The light from the sun shone on her way, and the waves gently rocked her as she managed to stay afloat.

From the Isle of Templos above them, she heard distant cheers and clapping. They had made it!

Exa smiled. Life is good at eighteen! She looked around for her twin and Violet.

Violet broke into the surface, coughing. Exa laughed, swimming to her.

Violet continued to cough and spurt out seawater. "That—was—intense!" she exclaimed in between coughs.

Exa laughed.

Violet rubbed her eyes and looked around. Then she stared directly at Exa's face.

"EXA!" she gasped. Her eyes widened in horror.

Exa's smile suddenly vanished. Violet was pointing at her face.

"Green and violet eyes," Violet whispered.

Chapter 3

Alec and Exa July

Because you have done this, cursed are you more than all cattle, and more than every beast of the field; On your belly you will go, and dust you will eat all the days of your life.

—Genesis 3:14

Exa quickly swam and swatted Violet's pointed finger. "Shh!" Her eyes were wide in terror.

Violet suddenly looked confused, and she glanced at the boy behind Exa.

Exa made the mistake of quickly turning around to see Theos, who instantly spotted what was amiss.

"How?" he gasped. "It's…unnatural. Like flying."

At that moment, Alec resurfaced, gasping. "I can't believe we did it!"

Theos turned to him. His eyebrows knitted, looking intently over his face "But your eyes are the same! Dark blue!"

A wave of horror suddenly washed over Alec, who was briefly startled by this observation. Alec casted a glance at his twin, and his fears were confirmed.

"How?" Theos repeated. "How did Eden allow this? How did AI allow this?"

Alec cleared his throat and quickly regained composure.

He calmly made his way to Exa. "Well, how did Eden allow flying? When did Eden allow flying?"

Theos sighed. "Again, it is a gift from Luna. Would you say that your eyes were a gift?" He turned to Exa.

Alec and Exa exchanged a look. "From birth, yes." Exa raised her chin up.

Theos's voice was rising. "Well, then you're an—"

"But Eden prescribed her with AI-grade lenses," Alec said, cutting Theos off. His voice was calm and cold. "The lenses must've gotten lost during the dive." He looked at Exa.

Exa gave a little smile. "I forgot to close my eyes."

"Still," Theos pressed. "How?"

Alec flashed him a look. "How can we question Eden? Did you question your goddess when you flew?"

Theos finally became silent. The salty air was then filled with tension.

Behind the twins, Violet was quietly observing and listening.

"If you're really curious, I was taken into intensive neonatal care upon our birth, then I was prescribed my lenses." Exa shrugged.

"I guess it's not every day you see curious things like this." Alec sighed and smiled, looking from Theos to Violet. His tone suddenly light and dismissive.

Theos gave a sigh of relief. "You're right, you're right. I'm sorry for being rude." Then he gave a loud snort. "And from someone who has been flying!" The twins gave a short forced laugh.

Violet was still silently frowning. Exa smiled and looked

at her. "I'll see you later, okay?"

"Later?" Violet repeated loudly, suddenly confused.

"*Voguer W*, all right?" Alec said.

"The ship?" Theos and Violet asked incredulously in unison.

"Party's boring there," Theos said, rolling his eyes. "You would want the booze and all the good stuff here!"

"Well, sure." Alec absentmindedly nodded. "My sister and I will just have to attend to this peculiar concern." He smiled at Violet. He was trying to be reassuring, but from the look on Violet's face, he knew he failed.

The twins raised their wristbands in unison. *ZAP!*

Theos and Violet found themselves immediately deserted in the middle of the Sophronia Ocean, staring at the spot the twins had been, the water still rippling in their wake.

AP!

The twins stumbled back into their bedroom—a big white ar room with two gigantic white round beds and a ing steel bookshelf dividing the room. Water dripped he floor.

xa fell to the side of her bed. She gave an almost hed cry. "You think he bought it?"

Ie has to," Alec sighed, pacing the room. "He has to."

Ve had this story prepared for the longest time," Exa lut this is the first time we had to use it."

ec ignored her. "He has to believe us. Theos has broken two Eden rules since last year."

"Eden's general rule, Alec! All that is good is made through artificial intelligence," Exa reminded him. "AI."

Alec fell silent, continuously pacing.

Exa continued, "You thought of it before in Templos, Alec. Flying can be done through AI's loopholes. Just tweak the gravitational points—"

"That would be floating," Alec suddenly said. "Yes, I have considered that. But Theos can direct his flight. It's him, not Eden."

"So you believe it's a gift from the moon goddess?" Exa scoffed.

"Of course not," Alec said. "But Eden is ever-changing, ever-revolving." He looked at Exa.

"And I'm a mutation," Exa sighed. "An anomaly."

"Don't say that," Alec simply said.

"No, because if there's something wrong with my genes, there's a bigger mutation inside genotypically," Exa said, making a face. "It must be in yours too." She shrugged, standing up.

Alec nodded. "We share this fate."

Exa sighed and looked as if she wanted to hug her brother, then decided against it.

"I don't think Violet is buying this, though," she said, walking to their shared bathroom.

"She's our longtime friend," Alec said. "She's smart, and she can be trusted."

"I hope so," Exa sighed.

She opened the cupboard and revealed hundreds of boxes of contact lenses. Their parents had explained that these lenses were not, as matter of fact, AI-graded, but good old Earth contact lenses. Decrepit. Two pieces of colored plastic lenses with no grade, made to change her eye color. It was necessary, their parents explained, so no AI could track the lenses. AI could detect another AI. A good old-fashioned trick would be the key.

Exa cleaned a brand-new pair of dark blue contact lenses. With one finger, she scooped one out and placed it over her eye. She repeated this with the other lens.

"Should we tell Mom and Dad?" Alec said.

"What for?" Exa said, blinking and then closing the cupboard, revealing a good old-fashioned mirror with her reflection now sporting two dark blue eyes. The anomaly had been covered. "I thought the crisis was averted."

Alec nodded silently, hoping that this was indeed the case.

Chapter 4

What Blossomed Under the Moon

I know a bank where the wild thyme blows,
Where oxlips and the nodding violet grows,
Quite overcanopied with luscious woodbine,
With sweet musk-roses and with eglantine.

— William Shakespeare
A Midsummer Night's Dream, Act 2 Scene 1

Violet felt the breeze and the tiny whirring zap that came with it when the twins teleported to the dock of the *Voguer W*. It was midafternoon. There was nonstop partying on board. She herself was drinking a bottle of gin-and-vodka solution, leaning on the white rails, enjoying the view of the Sophronia Ocean, and spotting blue-tipped dolphins that mimicked a siren's song from time to time.

She could spot the Island of Templos from a distance and tiny people cliff-diving and teleporting themselves. At least twice, she thought she spotted Theos flying around the island.

Flying, she thought, and heterochromia. Things whose given explanations didn't make sense to her.

"Violet!" she heard Exa call, but she refused to look.

Then she felt a warm hand on her wet shoulder. Alec.

"Hey, you enjoying the party?" Alec said, going to her side. Exa went to her other side. They were smiling at her. As if nothing had happened.

Violet scoffed and rolled her eyes. "How long?" She turned to Exa. "Since birth, yes, but were you planning to tell me?"

Exa blinked innocently and shrugged, glancing at Alec. "It never came up."

"Really?" Violet said wryly. "That you have AI-graded lenses?"

"We never really thought about it," Alec supplemented. "It's just part of our lives since then."

"What are you drinking?" Exa glanced at Violet's bottle. "Where can we get ours?"

Exa was trying to change the topic. Nevertheless, Violet shrugged and pointed toward the center of the dock, where rowdy boys were spilling alcohol from kegs.

"Oh, goodie. You want one?" Exa asked Alec.

Alec shook his head. Exa went ahead, leaving her brother with their cynical bestfriend.

"I'm sorry," Alec sighed, looking at Violet. "I know it must have been quite a shock."

"I've known you for years!" Violet said, exasperated. "Babies! How come I have never seen your eyes turn green and violet until now?"

Alec shrugged. Violet wanted to shout at him. Instead, she peered into his face. "Do you also have green and violet eyes?"

Alec laughed. "Oh, no, just Exa."

"It must not be a dominant gene," Violet said loudly to

herself.

Alec laughed again. "That's why you're in a mood. You're trying to figure it out."

Violet pouted. "Why do I feel like you're not telling me something?"

Alec laughed again and ruffled her black hair. "Silly! What's more to tell?" Then he leaned on the railing with her and looked up at the pink moon, which turned a bit brighter, almost colliding with the sun, which was nearly ready to set. "It's beautiful." The skies were starting to dim in the background, creating a beautiful spotlight.

Violet nodded absentmindedly.

"It's starting to be…you," Alec whispered.

Violet looked over curiously at him.

"It will turn violet in a while." Alec smiled, looking back at her.

Violet's breath caught. The pink light illuminated Alec's face, making it glow.

At that moment, Exa came back. "Man! Those drinks were too much!" She groaned.

Alec chuckled. "What are they up to?"

"Saw some kids from school doing silly games and dunking beer," Exa snorted.

"You bored?" Violet asked.

Exa shrugged. "Meh."

"I saw Theos flying over there a couple of times," Violet said, raising an eyebrow at Exa.

Exa and Alec exchanged a short glance. "He's probably

busy showing off," Exa chuckled. "But he knows where to find me if he wants to."

Alec scoffed. "He's not that impressive."

Violet and Exa chuckled at each other. "It's just a little crush," Violet told Alec. "It's harmless." She took a gulp of her drink.

Exa tugged at Violet's arm. "Did you see his six-pack?"

Violet almost snorted out the alcohol, burning her nostrils. "Exa!" She laughed.

Alec rolled his eyes. "You know that can be AI-generated, right? The aesthetic artificial intelligence centers can give you a six-pack within minutes."

"He looks rich." Violet took another gulp and laughed.

"He must be," Exa said, nodding. "Receiving a gift from the goddess of the moon."

Violet laughed. "You believe that?"

Exa shrugged. "I don't know what to believe," she lied. Alec stared at the skies.

"I can do research on that," Violet said.

The three friends looked over the horizon. The sun was almost setting.

"How about your heterochromia?" Violet suddenly asked.

Exa and Alec were startled at their friend's terminology. Exa thought of feigning confusion over it. Mutations were sporadic, unheard of, and *almost* at zero chances among human beings in Eden. Ordinary people wouldn't know the correct technical term. But she decided against it. She couldn't fool Violet, the most brilliant girl in their school.

Exa shrugged. "What about it?"

"Don't you want to know anything about it?" Violet asked her.

Exa smiled sweetly. "Eden did it for me when I was born. It's okay. My parents got the medical report." She suddenly hugged Violet, and she felt her friend relax. "Thank you so much, Violet. You always look out for Alec and me. Truly, it's nothing! I'm fine!"

"But a while ago—"

"It was the first time someone noticed them," Exa admitted. "It was unnerving. Honestly, it didn't feel right being judged based on your eye color." She pulled back from the hug and looked pointedly at Violet.

Violet bit her lip. "I'm sorry."

Exa smiled and let go of her friend. "It's okay. I'd be surprised if I were in your place." She and Alec changed a glance.

Alec looked slightly relieved. Exa seemed to have subdued Violet for now.

At that moment, they heard several zaps. They looked behind them. Some of the crowd from the Isle of Templos, including Rufus, Theos, Sandra, and Freia, had teleported to the *Voguer W*.

"*VOGUER W*!" Rufus shouted, brandishing a bottle of alcohol in the air. "LET'S PARTY!"

The ship roared and hooted in agreement.

"HAPPY BIRTHDAY TO ALL LEGAL ADULTS!" Rufus shouted again. There was another hoot.

"Happy birthday to us." Exa smiled at her brother and walked over to give him a small hug. Alec sighed and hugged her back. He kissed her forehead.

It was so sweet that Violet was inclined to look away. She cleared her throat. "Happy birthday. I hope you two enjoyed the day. It was fun, still, teleporting midair."

The twins laughed.

"It was thrilling!" Alec agreed.

"I'm actually glad we did it." Exa drank from her cup.

Theos suddenly walked up to them. "There you are!" He stared at Exa's face.

Alec and Exa smiled at him, trying not to look unnerved. "Theos!" Exa greeted. "Were you looking for us?"

"For you," Theos said, gazing at Exa.

Violet had to bite a laugh down. Exa blushed, almost as pink as the skies. Alec rolled his eyes and looked away, leaning on the ship's rails again.

"Would you like to dance with me?" Theos suddenly asked, offering her an arm and gesturing to the mad dance floor. The other teenagers were wildly singing and dancing to the latest beats, with reckless abandon. Their bodies swayed and grooved, with their drinks in their hands.

Exa laughed. "Sure!" She immediately took Theos's arm.

Violet looked around and spotted the red-haired lady and the brunette ones. "The girls from a while ago are looking for you," she told Alec.

Alec snorted. "Don't look, and scoot over closer to me."

This time, it was Violet's turn to turn pink. She looked away slightly as she stepped closer to Alec. His body was warm.

"They seem too old for me." Alec let out a tiny snort.

Violet laughed.

"Do you want to swim?" Alec asked. "I want to escape those girls for a while."

Violet nodded. "Oh, I've already gotten your bag from Templos." She gestured to Alec's bag on the floor. Alec thanked her, and took out goggles for him and Violet.

They placed them on. The goggles were light and flat. Violet turned on a small button on the side, and holographic text appeared across her vision.

"Why do you need the Identifier?" Alec asked, referring to the holographs on Violet's goggles.

Violet shrugged. "Well, we haven't swum these depths before. We could see interesting stuff."

"Like what?"

Violet sighed. "You know, Eden actually allows several mutations…"

Alec stiffened momentarily.

"…like the glowing algae and seaweeds of the ocean," Violet said. "The blue-tipped dolphins that can mimic human words, the spurting starfish, orchids, and violets submerged in saltwater, the gigantic mushrooms of the jungles. So many, Alec."

Alec nodded as he prepared to swim. "They are mutations."

"None of them are found in humans, though," Violet said.

Alec didn't answer.

"Let's go." Violet started climbing the rails, then dived into the ocean.

Alec followed suit. So Violet remained suspicious of them. Still, he expected she would not rat them out.

The underwater landscape was even more stunning. There were different kinds of fish. Violet was right to turn on the Identifier. She had already turned on her microphone and audio. Alec turned on his own and connected with Violet's.

"There!" Violet said, pointing to several schools of fish. Some were glowing. Some were translucent. Some were sparkling.

Alec couldn't believe his eyes. "I didn't know they existed." He turned on his own Identifier, and holographic texts appeared beside the fish he was focusing on.

It was a rhinestone fish.

"Would you believe there're more of them in the deep sea?" Violet said.

"Oh, really? That's amazing!" Alec said, admiring the different gemstones that served as the small fish's scales. It swam so slowly, as if it found the stones on its body heavy.

"It's a mutation too," Violet said. "Too much calcium and minerals. The rhinestone fish spends its days in the dark or buried in the sand. They have to procreate before their whole body is covered with rhinestones. It's their only chance. Once every surface, including their lips, is covered, the rhinestone fish die of hunger. But taking a rhinestone is dangerous because their flesh will produce poison when these stones are plucked."

Alec listened and watched the rhinestone fish struggle to get out of their way. Then he spotted flowers.

"Oh, saltwater violets!" He pointed. There were purple buds among the floating seaweeds.

"It's almost nighttime," Violet noted, watching the saltwater violets. "Their petals open just in time for sunset. The petals look thin, but they're actually sturdy. They elongate for a few centimeters, then glow."

As if on cue, the saltwater violets bloomed.

"Let's go up!" Alec suddenly grabbed Violet's hand, and they immediately swam up to see what was left of the sun disappear into the horizon.

The skies were dark, and the stars twinkled brightly. The moon glowed brighter pink. As the night went on, it would grow even brighter. Because of the saltwater violets, several areas of the ocean glowed light to dark violet.

"Beautiful," Violet gasped as she watched the magical setting unfold.

"It is," Alec said.

Violet turned to him to see him looking intently at her. She suddenly realized how close his body was to hers. He hadn't let go of her hand. Slowly, Alec moved his face closer to hers.

Should she close her eyes? Violet thought. She could feel her heart hammering in her chest, and she was sure Alec could hear—

"ALEC! THERE YOU ARE!" Freia's shrill voice broke the moment.

SPLASH!

Alec and Violet looked. The two girls from the Isle of Templos had dived from the ship and were swimming together.

"So much for hiding," Alec chuckled. He waved and smiled at the two girls.

Freia immediately reached them and threw her arms around Alec. "Hey, I've missed you!" she crooned.

Violet wanted to gag. She noticed that Freia had changed to an even skimpier swimsuit, leaving nothing to the imagination.

Alec smiled and placed an arm around Freia's waist. Violet's eyes narrowed. "We just went for a dive."

Freia and Sandra didn't glance at Violet. Violet moved away a little.

"What did you think of Theos?" Sandra asked no one in particular. "He's amazing, right?"

"Oh, yeah, I heard he's playing with your sister on the deck," Freia said to Alec. "Come, let's join them!"

They pulled him toward the ship. Alec turned to Violet. "You coming, Violet?"

Violet smiled sadly and shook her head. "I'll stay here for a little bit." She thought she saw Freia smirk at her. She watched Alec swim away with the two girls toward the *Voguer W*.

Then she was alone in the cold ocean, with the bright pink moon and glowing saltwater violets.

Chapter 5

The Moon Priestess

At last the day came for the naming of their new-found child, so the old couple called in a celebrated name-giver, and he gave her the name of Princess Moonlight, because her body gave forth so much bright light that she might have been a daughter of the Moon God.

—Yei Theodora Ozaki,
The Bamboo Cutter and the Moon Child

Alec and the girls were dripping wet when they returned to the ship.

"Where's Exa?" Alec immediately asked.

Rufus answered, "She and Theos took off just now. Theos was saying he'll show her something." He took a gulp of his drink.

Alec's eyes narrowed, and he inwardly berated his sister. Beside him, the two ladies crooned and giggled. "Oh, I guess they're getting some action."

Alec ignored them. He looked up to the skies, scanning the expanse, hoping to see two figures flying and soaring. He sighed when he couldn't. This day has seen enough action. He fervently hoped no accidents would follow.

Exa couldn't believe it. She should close her eyes; the contact lenses might fall off or something. But she couldn't. She was flying!

The dark blue sky stretched around like a satin blanket around them, studded with twinkling stars. And the moon—how glorious was its pink glow! She swooned, mesmerized at the beauty around her. The night sky could envelop their bodies in its cloak, and she wouldn't mind.

She found herself tightly clutching Theos's bare back, her arms around his neck and shoulders, and her legs around his waist. Theos was flying higher and higher and further until the *Voguer W* was just a tiny dot below.

All around, she could see the hundreds of Eden's floating islands, each with its unique beauty. Eden was indeed a paradise. Eden in its sprawling splendor! How breathtaking!

Thanks to AI, she had to remind herself. However, that sent a bitter feeling down her stomach, knowing there was a big chance she didn't belong to paradise. Her smile slowly faltered.

"Is Isle Luna near yet?" she called loudly to Theos, against the howling gale. The wind current was getting stronger as they flew higher.

"There!" Theos pointed. "It's one of the highest floating islands." Isle Luna bathed in silver and pink moonlight. It had a mountain towering above the clouds, and Exa couldn't see its peak.

Through the clouds, through the blanket of stars, Theos flew even higher. The air started to get thinner. Exa reminded herself to slowly breathe.

Theos flew over the edge of Isle Luna, which was surrounded by Ficus trees, whose thick vines and roots intertwined, entangled like ancient serpents covering the earth. He slowed down upon reaching the edge and gently set her on the ground.

Exa grinned, looking around. This was her first time on Isle Luna. The moon here looked even more prominent. She looked at the mysterious Ficus trees around them.

"Have you ever played with Ficus trees?" Theos asked.

"Can't say I have," Exa admitted.

The two went over to the nearest tree. "Its roots and vines have the memory of touch," Theos explained. He started hopping from one root to another, then holding vine after vine. To Exa's surprise, the roots and vines began to produce glowing yellow dots, one after the other, like a piano whose keys lit up on touch. Theos stepped away.

He looked at Exa. "Can you remember the sequence of vines and roots I've touched?"

Exa giggled at him. It was child's play. She jumped from one root to another, holding vine after vine, mimicking Theos. Suddenly, the Ficus tree shook, and the vines and roots started moving, creating an entrance for them.

"Whoa," Exa gasped in amazement.

"See?" Theos said. "Ficus trees are one of the best hiding places. In some of the biggest and oldest Ficus trees here, some people made them their tiny houses. You can fit a lot inside."

"So cool!" Exa exclaimed. "We don't have any Ficus trees back home."

"Only a few islands have them," Theos said, gesturing to the empty black space the Ficus tree was offering to them. Exa stared.

"Do you wanna go inside?" Theos asked.

Exa was unsure. It looked very dark. She looked at Theos in confusion.

Suddenly, a voice sounded from the hollow of the tree. "Theos? Is that you?" It was a woman.

Theos beamed. "Oh, hey, Mom!"

Mom? Exa peered more carefully inside. And there she saw a faint orange light ahead, growing nearer and bigger.

"Yeah, Ficus trees can serve as passageways as well. This one's AI and not a mutation like the memory thing."

There it is, Exa thought. That word again. Mutation.

"Eden was able to bend space and time inside Ficus trees," Theos explained. "Amazing what AI and nature can do."

Exa politely smiled and nodded, keeping her eyes on the orange light approaching them.

Then she saw a woman's face, illuminated now by the pink moonlight. Her face had the light wrinkles of a fifty-year-old. Most opted to go with AI aesthetics, removing all wrinkles and unwanted fats and, as Alec pointed out, getting six-pack abs.

Theos's mom, despite her age, looked ethereally beautiful. Her skin was dark like Theos's. Her hair was pure silvery white, straight and long, dancing with the breeze. She was wearing white flowy garments made of silk and adorned with a big pin.

Exa stared at the pin. It was in the shape of a golden apple with the words "The Tree of Life."

They were right. Not only was Theos rich, his family was affluent. His mother belonged to the Tree of Life Council. The elite politicians of Eden 2.0. Maybe that was how Theos got away with his antics.

"Welcome!" The woman suddenly hugged Exa. Her

movements were light and graceful as if she were dancing. "Isle Luna welcomes you." Her voice was sure but calming.

Theos smiled. "This is Exa July. She turned eighteen today. We met at the Sophronia Ocean. She saw me fly."

The woman tsked. "I hope you weren't showing off."

Theos winked. "Just a bit." Then he turned to Exa. "My mom is the high priestess of the moon. She asked the moon goddess to grant me the gift of flight."

The woman turned to Exa. "My name is Levida Luna. After the moon, of course. The goddess is generous to those who ask her." She gestured at the pink moon.

"How did you ask the moon goddess?" Exa couldn't help asking.

"Would you like to know?" Levida asked. "Follow us. The Ficus trees lead us to our temple." With that, Levida entered the dark hole of the Ficus tree. She raised her wrist, and only then did Exa realize that the orange light was from the woman's wristband.

"After you." Theos gestured, guiding Exa in.

Exa hopped into the hole. To her surprise, it was bigger inside than it looked from the outside.

Theos and Exa pressed their wristbands, and orange holographic orbs appeared, lighting their way. They followed Levida, and only after a few steps did Exa watch the end open into a moonlit room.

They stepped out of the Ficus tree. They found themselves inside the temple, where the roots of Ficus trees wrapped themselves in the temple's humongous circular columns. There were no walls. Instead, they had the largest view of the pink-cratered moon of Eden.

Exa noticed a huge revolving mirror on the ceiling and a circular altar in the center of the room.

"Would you like to see?" Levida asked. "You're just in time for the evening ceremony."

As if on cue, several roots of the Ficus trees slid and moved to open many holes, from which men and women in white garments emerged.

"This is the evening ceremony," Theos said, smiling at Exa.

Exa felt the hair on her arms rise. She stilled herself. She was completely unfamiliar with Isle Luna's paganistic rituals. If she ever felt the need to escape, she might be able to quickly teleport. Her eyes wandered to Levida's golden apple pin.

The people were silent as they gathered at the circular altar. Levida moved beside the altar and her son.

Theos squeezed Exa's hand in reassurance, smiled, and then left Exa to stand beside his mom.

"Dearest moonfolk, we have a visitor who will observe our evening ceremony," Levida called.

"Welcome in the name of Luna," the people said in unison.

Exa tried to smile, but to be honest, she was completely weirded out. That was not something one could say of a date. If this was one.

"Tonight," Levida said in a loud voice, "we continue to ask the goddess for her continuous blessings, that we may share in her glory and power. That we may be bestowed more gifts from the moon." She signaled to her son.

Theos climbed up the altar and opened his palms wide. He closed his eyes and lifted his chest to the moon.

The people also closed their eyes, held each other's hands, and began chanting in a language Exa had never heard of in her entire life. Exa saw Levida turn to a computer-like machine beside the altar and pull a lever. It made the giant revolving mirror on the ceiling turn and twist many times until it reflected the pink light of the moon and, with another twist, sent this reflection to the altar, to Theos.

Levida joined the group in chanting. Exa noticed that she was fixing things on the machine as she did so. Exa briefly saw a hologram with millions of codes. Then it was gone.

AI, Exa thought. She looked up and squinted toward the moon. She couldn't see anything more than a giant mirror reflecting the moon. But it felt…unreal. The light felt unreal. Off, somehow.

She looked at the machine near the altar. What was it for?

"I CAN FEEL IT! I CAN FEEL IT!" Theos suddenly shouted. Then he screamed.

Exa looked in alarm. The people's chants grew even louder and stronger.

Theos shouted for a long time, then when he stopped, the chanting slowed and became quieter. The pink light disappeared.

A hologram, Exa thought. Or a field ray of light. The pink glow wasn't just from the moon. There was an AI guiding it.

The people bowed, including Theos and Levida.

"We thank you, moon goddess!" Levida called loudly. That instant, Theos flew above the altar.

Everyone cheered while Exa looked on silently. Her mouth gaped at what she had just witnessed.

It wasn't the moon, she realized.

Later that night, Theos came home after flying Exa back to the *Voguer W*. Alec had been visibly irritated and glared at Theos; he had waited a long time.

And now, to his surprise, another figure was waiting for him back in the temple: his mother.

"You date so many girls, Theos," she said without a hint of humor in her voice. "Why bring that one?"

Theos shrugged. "I thought she'd be interested in our culture."

"Ah yes," Levida said. "But not all can join the Isle. Why this particular girl, though? A lot have pestered you before to take them here."

"I think she's gifted by Eden herself," Theos said.

"By Gaia?"

"Not Earth," Theos scoffed. "Eden. The girl has two different eye colors."

Chapter 6

The Tree of Life Council

The Lord God made all kinds of trees grow out of the ground—trees that were pleasing to the eye and good for food. In the middle of the garden were the tree of life and the tree of the knowledge of good and evil.

—Genesis 2:9

Levida Luna stared after the retreating figure of his son. It was almost dawn. But she needed to send a message first.

The pink glow of the moon was waning. Another day was breaking. Indeed, another day. Levida sighed as she mulled over what her son had told her.

She lifted her wristband and sent a voice message to the other twenty-three members of the Tree of Life Council.

"Activating code 3 and code 4. Please acknowledge the urgency of the situation. I call for an emergency meeting at the soonest possible hour," she announced and pressed send.

All throughout Eden, twenty-three wristbands lit up in red.

The Tree of Life Council consisted of twenty-four members.

They were in charge of all the rules and regulations of Eden 2.0. They were the fiber of what made Eden 2.0 a paradise. Their eligibility as council members was based on merit, intellectual ability, and, most importantly, ancestral history. Every member of the council came from parents or grandparents who had served Eden 2.0 since the beginning.

The council decided what rules to follow, what rules could be bent, how to use AI for the welfare of the people of Eden 2.0, and how to make life in Eden as it was now.

Levida had only slept for two hours before she made her way to the Tree of Life. It was located in the center of Eden, one of the very few islands resting on Eden's crust amid the hundreds of floating ones.

Just behind the Tree of Life was the Tree of Knowledge of Good and Evil.

ZAP!

Levida teleported herself in front of the two structures, the symbols of power in Eden 2.0. She took a deep sigh and admired them.

The Tree of Life was a blue glass building situated inside a massive tree in the middle of the desert, while the Tree of Knowledge of Good and Evil was a building of glass spheres situated inside another tree that bore gold apples.

For security measures, they had forbidden teleportation inside the two buildings. Levida had no choice but to take the long staircase to the entrance. There were no guards in sight, but the big holographic doorway began to scan her. It caught sight of her golden apple pin. Immediately, she was granted access.

Inside, the council was ready to convene. The members all gathered at a circular golden table. The floor in the center was empty except for a glowing blue pool of light.

"I thought first light," one of the members, Rosarius Santi, said. He was the most conservative member of the council. "Yet you are the last to arrive, Levida."

He was also one of Levida's strongest oppositions in the council.

"Oh, Rosarius, you know I feel strongest during the night," Levida quipped.

Rosarius scoffed. "Perhaps paganism isn't the way to live if you have to fear the day."

Levida snapped at him, "Our religion is of the purest utmost—"

BANG! BANG!

The gavel sounded. It was Agham Malaya, the president of the council. Of course, he was in the council by birthright. He was also found to be intelligent and had good academic merits. The majority of the council members had elected him president.

"ENOUGH!" he bellowed. "The urgent matter at hand is NOT the place of religion in Eden, for such will take a long time to settle. The matter is of codes 3 and 4, which YOU"—he pointed straight to Levida—"have called."

Levida went to her chair but didn't sit. She looked at the other members slowly and warily. Twenty-four people were so heavily distanced in this room that one had to shout.

"Everyone," she called in a loud voice. "I have heard of a report of Subject Code 3. A girl who just turned eighteen yesterday. A case of heterochromia."

There were several murmurs.

"Code 3," Agham began, "as you all know, was the unsolved case of a heterochromia twin who never made it to Intensive Neonatal Anomaly Care. Instead, we believe it was stolen by its parents, WHO ARE STILL ON THE RUN." His voice became louder at the end, emphasizing the gravity of the situation.

The glowing light in the center flashed a holographic screen filled with details of code 3. It showed the date: January 31, 722 AE. Underneath was a picture of an infant with unmistakable purple and green eyes. Then Agham changed it to a recorded video scene: a naked holographic woman in a bathtub who refused to give the infant to the AI doctor. The woman started shouting and screaming. Another holographic figure—a man—suddenly grabbed a chair and brought it down on the AI doctor, killing it. Out from the doctor's head came several wires and bursts of electricity.

There were several murmurs inside the chamber, with some members pointing and shaking at the video.

"Mutations are never allowed in Eden," Levida agreed in a loud voice.

"It looks more than a coincidence. I have to agree," another member, Edmund Siegart, said. He was one of the richest council members, if not the most. He had personally financed several projects of Eden 2.0. "If this girl is indeed Subject Code 3, where is her twin?"

Levida looked down. "I haven't seen him."

"You called us for nothing?" Rosarius scoffed. "Levida, is it possible your source is lying?"

Levida's eyes shot daggers at him. "The girl was celebrating her eighteenth year at the Sophronia Ocean yesterday—"

"Where hundreds of coming-of-age teenagers go," another member, the youngest member, Jet North, said. He was only nineteen years old. His father, Jet Senior, had died the year prior after living 467 years in Eden. "I have to agree with Rosarius. How sure are we that this is Subject Code 3?"

"Did you personally see her?" Agham demanded of Levida.

Levida squared her shoulders and nodded her head. She anticipated the next question.

"Were her eyes of different colors?" Agham continued.

"No," Levida said quietly. The word echoed around the chamber, prompting a wave of irritated murmurs and exclamations from the members.

"Then what is our basis?" Rosarius cried. "Where is the heterochromia?"

"My source says she has green and purple eyes!" Levida said loudly. "Remember that we have never disclosed to anyone outside the council the eye colors of Subject Code 3."

The members fell silent.

"Green is the rarest color, and complete heterochromia is the rarest type," Levida continued. "When I met the girl in Isle Luna, she had dark blue eyes."

"How was that possible? A trickery? Graded lenses? Isle Luna has security measures, doesn't it?" Jet asked. "Can we identify her based on the retina records? Is it possible?"

"I already checked," Levida said. "She was the only one unaccounted for."

There were several gasps.

"How can AI not detect—"

"Is this really Subject Code 3–"

"It must then be an Earth technology," Rosarius said, looking at Agham. "If AI cameras weren't able to identify her."

"If that was possible," another member began, "why haven't we checked every camera on every island?"

Several members groaned in irritation, including Levida and Rosarius. Jet seemed to snigger quietly.

Agham narrowed his eyes. "Simply because we did not put AI cameras on every island. Instead, we opted for a better way: a wristband for every human in Eden."

"Her wristband!" Rosarius exclaimed. "Does she have one?"

Levida nodded. "She has."

There were several murmurs. "Then she has indeed infiltrated the system! Is that possible?" someone exclaimed.

"Which brings us to code 4," Agham said. "The hypothesis that these twins have infiltrated the system."

The council members looked at each other and started nodding in agreement.

Agham looked at Levida and loudly asked, "May we know the name of your source?"

Levida suddenly looked down. She had known that there was a possibility Agham would ask. She sighed and stammered a little. "My son, Theos."

The murmurs among the council members grew louder.

"And how, pray tell, did Theos come across the heterochromia twin?" Rosarius Santi suddenly laughed, suspicious.

Levida shot a look at Rosarius. "He met her at a party in the *Voguer W*."

Rosarius scoffed. "When will you train your son? He's

enjoying life too much! Partying too much—"

"How I raise my son is no concern of yours!" Levida suddenly shouted.

Agham cleared his throat and shouted, "Enough! Let's get to the subject at hand! We have to launch a full investigation into the girl, her family, and her friends."

"Those in favor," Agham called, "scan your gold pins."

In the center of the room, the blue light counted twenty-four gold pins.

Chapter 7

Not the Moon, Just AI

When religion and politics travel in the same cart, the riders believe nothing can stand in their way.

—Frank Herbert, *Dune*

Violet woke up early the next morning despite having partied out late, thanks to her father's wristband alarm. Her mother was also called to work early.

"Security measures." Her mother, Azula, smiled warily and kissed Violet before leaving at first light. Her mother was one of the scientists at the Tree of Knowledge of Good and Evil. "Please don't be late for school. Your exams are coming soon."

As an only child, and with a household AI to command, Violet was left alone. And she was worried. What had caused her parents to wake up and go to work so early?

Security measures? There were very few security breaches in Eden. She thought of Theos taking off his wristband. Removing wristbands could be punished with temporary withdrawal of AI benefits, such as teleportation, for a week.

She got up after a few minutes and went over to the kitchen.

"Pancakes for breakfast," she said loudly.

"Gladly, Violet." An AI housemaid, almost lifelike, came to life and started cooking. After a few minutes, Violet began eating.

She thought of last night bitterly. Exa had returned to the *Voguer W* very late. Alec was obviously irritated, and the girls didn't leave his side. Violet had to endure Rufus's company. It was awful. The twins even left without saying goodbye to her.

At seven, Violet enjoyed a nice warm bath. She had an indoor pool all to herself.

She even had an AI designer robot to pick her clothes. Yes, her family was also one of the richest in Eden. Maybe that was why she wasn't too impressed with Theos yesterday.

At eight thirty, Violet chose their library, her favorite place, to have classes.

She adjusted the settings of her wristband and pressed. A holographic classroom came to life around her. At the back, she spotted the holographic image of Rufus, who was checking something stuck underneath his shoe. Violet rolled her eyes. Beside her, she saw the two holographic figures of Exa and Alec appear.

She looked at them, unsmiling.

Alec looked down. Exa sighed. "Violet, we're sorry to take off like that—"

"You do that a lot," Violet said accusingly. "Have you noticed?"

Exa and Alec exchanged glances. Violet could imagine them in their own house, in their own bedroom, sitting side by side, maybe on Exa's carpet.

"We're sorry," Alec repeated. "We were just worried that Dad would go berserk again…"

Mr. July had a tendency to panic, yes. "Not an excuse." Violet pursed her lips. "I had to talk to Rufus for hours!" She rolled her eyes. She turned to Exa. "And you! Where did you even go with Peter Pan?"

Exa giggled at the name. "You wouldn't guess!"

"Isle Luna?" Violet wearily said.

Exa nodded excitedly. "Yes! And you were right! It wasn't the moon!"

Violet wanted to roll her eyes again. Of course it wasn't the moon. Compared to Earth, Eden had its moon way closer to it, and thus, Eden's moon had higher radiation levels than Earth's moon. However, despite several research studies, there were no findings indicating that these radiations could cause full mutations, especially in humans, such as flying. Also, human genes were more challenging to manipulate than plants. Even animals had fewer mutations compared to plants.

Exa continued, "They have an altar where they made Theos stand. It was their nightly ceremony. There was a big mirror on the ceiling which reflected the moonlight onto Theos. Then, there was a computer-like machine. During the ceremony, it was on, and I swear, there was light coming from it. Like radiation."

Alec looked bored. Of course, he had already heard this story last night.

"Then Theos shouted that he could feel it," Exa said. "Then he flew."

Nothing in Exa's story exactly satisfied Violet's questions. A dubious machine? Who even had machines in this era? Everything was microchips and holograms. The power was in a flick, not in a lever pull.

"Oh, Theos's mom was the high priestess," Exa said.

"And a member of the Tree of Life Council."

Now, that was something. Violet's eyes widened.

Beside Exa, Alec sighed, looking glum.

Chapter 8

Under Attack

to believe in landscapes illuminated by a god bathed in blood,
yet to be born. The world wanted explanations:
why things exuding warmth burn

upon the slightest touch- singe a song
of pain on skin. People started believing
in what was bright, blinding: the sun –

—Alyza Taguilaso, *What Burns*

It was the middle of the school day. They were learning about pulsed energy projectiles in their fifth hour when they heard a huge commotion outside their room. Voices and the sound of breaking glass.

Exa and Alec exchanged glances.

"I'll take a look," Alec said, turning off his holographic image from the class and getting up from the floor.

Exa remained cross-legged on her carpet, taking notes. Her focus returned to the physics of pulsed energy.

The glass door to their bedroom automatically slid open. Alec went out. He glanced left and right. Their parents' room seemed to be untouched. And there was silence.

He sighed. He must be imagining things. For some reason, learning that Theos's mother was a member of the Tree of Life council unnerved him more than he wanted to admit.

He turned around to enter class again when—

PING! TSS!

A line of green laser shot across in his peripheral vision and hit the opposite wall, creating a burn mark on the previously immaculate white paint. A tiny wisp of smoke emerged from the mark.

Alec's eyes widened. Exa raised her head at the sound, alarmed.

There was another second of silence, as if time was moving slowly to allow them to process what was happening.

Then came the noise.

"RUN! ALEC! EXA!" Their mother's scream shattered the air.

Then, there were footsteps up the stairs.

CRASH! BANG!

"DO NOT USE THE WRISTBANDS!" their father shouted.

"You have to surrender under the code—" a robotic voice said.

Alec ran back to their bedroom. He grabbed Exa's sleeves. "TURN OFF!"

Exa immediately shut down her wristband. The holographic classroom disappeared into oblivion. "Should we get them off?"

Alec shushed her. "Not yet! Maybe! I don't really know!"

"We have to get underground!" Exa said.

Outside the room, there was more banging and sounds of lasers and tasers. They could hear their mother and father shouting and fighting back.

"We have to help them!" Exa exclaimed to Alec.

Alec did a double-take, looking to where their emergency exit was hidden, but he agreed with Exa. "Let's go!"

The two ran out of the room.

PING! PING! More green lasers bounced around the room.

"At this rate, the whole house could burn," Alec said as they tried to avoid the lasers.

"Let's throw furniture at them!" Exa said, grabbing a chair beside her. They went over to the staircase and immediately threw it below. It landed on an AI soldier, instantly beheading it and revealing a complicated circuitry inside.

"Grab the laser gun!" their father shouted. Their mother picked up the laser gun and began shooting the other soldiers.

"There're too many!" Alec shouted. There were even more outside their house. Their glass windows revealed an entire horde of AI soldiers forcing their way in. They were faceless, metallic robots. These soldiers were programmed to capture or kill. That was why Eden never bothered to put faces on them or even try to imitate human appearance or behavior. They blindly followed the law that was programmed.

They know, Alec and Exa realized in dread.

"We have to run!" Exa shouted, throwing different things at the AI soldiers while trying to duck.

Their father looked up at them. "NO!" he bellowed. "YOU TWO RUN!"

"What are you saying?" Exa cried out.

"We'll hold them back for you!" their mother shouted. Another AI soldier dropped, electrocuted by a laser beam.

Mr. July immediately got the gun. "ALEC!" he shouted, throwing the weapon toward him. Alec ran forward to catch it. As he did so, a green laser suddenly zapped past, grazing his shoulder.

"ARGH!" he shouted as he fell to the ground with the gun.

"ALEC!" Exa shouted. "YOU'RE BLEEDING!" She ran toward her twin. "Can you stand?"

In pain, Alec groaned and nodded. His shoulder hurt like hell, but he had to fight. He stood up. "LOOK OUT!" He shot an AI soldier trying to come up the stairs.

"RUN! RUN!" their mother shouted at them.

A green laser suddenly hit her on the back.

"ARGH!" she screamed, falling forward. She started to convulse.

"MOM!" Alec and Exa shouted.

"Nia!" their father screamed, running to his wife. He started shooting at the soldiers that began to surround them. He looked up at the twins and glared at them. "IF YOU VALUE US, YOU WILL RUN NOW! OR EVERYTHING WILL BE IN VAIN!"

Exa had tears running down her cheeks. "Dad, we can't leave you and Mom!"

"EXA!" Mr. July hollered, shooting the soldiers in manic mode.

Alec took a deep breath. His hands started to shake. He knew what must be done.

"ALEC!" Mr. July called. "GO!"

The twins looked outside the big glass windows. The AI soldiers hadn't even diminished in number. Even without calculating their chances, Alec knew they were defeated.

"We have to go," he told Exa quietly as he continued shooting the AI soldiers surrounding their parents.

"What?" Exa gasped. "But Mom and Dad!"

"We stay here, and no one will survive." Alec looked her dead in the eye. There was no expression on his face. "Dad is right."

Exa cried wordlessly. Then she shouted in anger as she hurled a table down below to several soldiers in a final blow.

With a last *PING* from his gun, Alec immediately grabbed his sister's hand, and they ran. Several soldiers saw and started running toward them.

"NO!" they heard their father cry. "FACE US INSTEAD!"

There were more green lasers. Alec and Exa kept ducking down to avoid them.

"The passage!" Alec pointed.

Their parents had built the emergency exit behind their family portrait when the twins were young and it had never been used until now. It was made solely for this event.

Exa swung the portrait, and the twins entered the dark tunnel.

Alec reached out to the wall and found the loose brick in the blackness. He pushed on it. Ten openings revealed themselves. He went to the seventh opening on their right.

"July," he muttered. It was their password.

Outside the portrait, the shouts of Harold and Nia July rang with the continuous zaps of laser beams.

Chapter 9

Gone

Some are born great, some achieve greatness…

—William Shakespeare, *Twelfth Night,* Act 2.5

Violet had noticed them instantly.

"The pulse length can be very short, resulting in very high peak powers with relatively low pulse energy…" the professor said when Violet saw Alec's holographic figure disappear with a soft zap.

It took only a minute, and then Exa's figure was gone.

What was that about? Violet thought. Even though the twins could be mysteriously weird, even leaving a party without saying goodbye, they wouldn't leave in the middle of a class.

Education was valued in Eden, and unlike on Earth, people could choose what subjects to study once they finished basic reading, math, and writing, as preparation for the careers they wanted to have in the future. Not that having a career was a must because almost everything in Eden was acceptable to everyone. A career was not a necessity for survival but for self-fulfillment.

The hour ended, and the twins didn't come back.

"Where are you? The class is over," Violet said in a voice message to the twins. Nothing. There was no reply.

There was also no indication on her wristband that the twins could hear her message. That was weird, Violet thought.

The next class began. This time, it was the course on Eden versus Earth. It was one of her favorite classes with Alec and Exa. There was no way the twins would miss this, especially since the class would be debating politics and government.

Violet sighed, growing impatient. She surrendered to the thought that the twins might not return for class.

After class, Violet entertained herself by playing chess with an AI. She won in the second round.

By dinnertime, her parents came home. She heard their loud voices at the door, and her father was almost shouting.

"—it is of the utmost importance!" he was saying. "We cannot let such an anomaly—"

"Remember that these are kids we know," her mother tried to interject.

"You know them—"

"I do!" her mother exclaimed. "These are your daughter's friends!"

Violet's ears perked up. She started making her way to her parents.

"They are under investigation! Their identities have cropped up!" her father said. "You even confirmed it—"

"But sending an AI army isn't the answer!"

"YOU WHAT?" Violet gasped loudly.

Her father, Agham Malaya, and her mother, Azula Malaya, turned to their daughter, whom they hadn't realized was nearby.

"Violet..." her mother began. She hadn't taken off her

white laboratory suit.

But Violet was looking at her dad. "Why did you do that? Sending an army? What for?"

Her dad was still in a black suit and cape and wearing a big golden apple pin. As the president of the Tree of Life Council, his pin was the biggest and most prominent.

Agham squared his shoulder. "Their company is not the one you should keep," he said strictly.

Violet was taken aback. Who was he to dictate? The president of the council. "What did they do wrong?"

"They are fugitives," her mother explained carefully. "Their identities cropped up when we were investigating a previous case."

"A previous case…?" Violet asked incredulously. "I'm almost always with them! They are not criminals!"

"We cannot disclose the case," Azula said. "You know the rules of the council."

"Maybe pass the exam, and we will tell you as part of your training," Agham said. He turned to his wife. "It's better to do this now while Violet hasn't started training."

"Agham," Azula sighed. "Have some compassion!"

Violet looked tearfully between her parents. What was going on?

Her father turned to her again. "You won't have to face your friends when the time comes. I will end the case now! I will make sure those twins and their parents will pay for their crimes!"

Chapter 10

Fugitives

The Lord God therefore banished him from the garden of Eden, to till the ground from which he had been taken.

—Genesis 3:23

They couldn't turn on their wristbands for light. They couldn't risk turning on a button despite their notifications telling them that Violet had sent them a voice message.

Exa and Alec, hand in hand, traced the dark passageway, using the wall as a guide.

Exa would occasionally sniff. Gone were the shouts of their parents and the sounds of laser guns. Sometimes, they heard rats scurrying.

They didn't know how long they'd been walking—it must have been hours. They didn't even know where they would end up. All they knew were the instructions their parents had left for them: seventh passageway, then continue on.

Several hours passed. The twins could feel their feet aching, but they didn't complain. They just continued walking. They couldn't see any light, still, but their eyes had adjusted to the darkness.

Above their heads, they heard rumbling.

"We must be below the city," Exa said, breaking the silence for the first time.

Alec nodded. They continued walking. Then he touched something cold and hard.

"Hold on," he said. "I feel something." He released Exa's hand and proceeded to feel the cold surface.

"What is it?" Exa said.

"It feels like a door," Alec said, pushing then grunting. "Look, here's a latch."

"I think we're meant to pull," Exa said.

"Okay, help me," Alec said. He felt Exa beside him as they held on to the latch and pulled.

CREAK! The door opened a little. A fresh cold breeze greeted them. The twins peeked around. It was another chamber with a ladder. Above, they could see the pink sky. It was nighttime already.

"More," Exa instructed as they tried their hardest to pull the door.

CREAAAAAAAK!

The door swung loudly. The hinges smelled of rust as if they had not been used for years. Alec and Exa stepped out onto the chamber. Underneath their shoes were muddy puddles.

"It rained," Alec merely said. He went to the ladder and grabbed a step. It was wet and sticky.

"Be careful," Exa said, going after him.

They started climbing up. Alec could see the giant stars and moon above. "We're here," he breathed as soon as his head was out.

"Exa," he called. "We're probably in a different area of the island."

"How about you step out?" Exa called from below.

"Sorry." Alec stepped out and offered a hand to his twin.

They found themselves on top of a hill overlooking the many houses on their island. And their home—it was burning. It was so easy to spot with the dark smoke and flames going up and up.

Exa whimpered, and the twins hugged each other, saying nothing. What was there to say now? Did their parents get out alive? Would they be able to see them? What would happen now? Where would they go?

Their wristbands suddenly vibrated and flashed red. It was an announcement from Eden. The twins glanced at their wristbands.

Throughout Eden, all wristbands flashed a holographic image of the twins.

"ALERT! FUGITIVES! WANTED FOR ESCAPING THE LAW!" a robotic voice announced. Below the line was the picture of the twins. "ANYONE CAUGHT TRYING TO HIDE THEM WILL BE DEALT WITH ACCORDINGLY!"

Alec and Exa looked at each other grimly.

"They can track us with these," Alec said. "We should have removed them earlier." He removed his wristband. Exa did the same. Then they both dropped them into the tunnel they had just come from.

"I guess it's the fugitive life for us. Where do we go?" Exa said. Alec shrugged.

They had no one and nowhere to go. "It's good you were able to bring a bag," Alec sighed.

Exa frowned, looking at the small backpack she had grabbed. It was their emergency bag, which was always underneath her other bag. "A change of clothing for you and me. That's it."

Alec wanted to laugh. He hugged his twin again. "Fugitives," he repeated, smirking. "We need a plan."

Exa nodded, looking down at the wristbands they had just disposed of. "We have to move."

Chapter 11

The Next Council Members

...and some have greatness thrust upon them

—William Shakespeare
Twelfth Night, Act 2.5

"Anyone caught trying to hide them will be dealt with accordingly!" The red hologram warning dissipated back into the wristband.

Violet stared at her wristband in silence. She was alone in her room. This was her father's doing. Now, the twins would be living as fugitives.

She sighed as she crossed the room. There must be something she could do, but what?

Going outside her bedroom, she heard her father's booming voice. "We don't have prisons, but please detain them in the interrogation chamber!"

Were the twins caught? So quickly? Violet tiptoed over the corner. Her father was leaning on the glass balcony, talking to someone on his wristband.

Her father paused, then sighed. "They have to tell us something. They've raised the twins for eighteen years!"

Mr. and Mrs. July, Violet realized. They were caught. She turned to her wristband for a virtual map. A hologram map of Eden appeared before her. She pushed the coordinates of the Julys' house.

She nearly gasped aloud. The big white glass house had been reduced to ashes, and there was almost nothing left. Robots would come any time now to clean up the mess. Where would the twins go?

Her wristband vibrated loudly. Violet saw her father start to turn at the sound. She immediately tiptoed her way back to her room before she was caught.

Who was this now? She looked at the message on her band. It was from Theos.

We need to talk.

Violet's eyes narrowed. Besides her, only Theos had seen Exa's eyes. He must have ratted her out to his mother. Or something else had happened besides the ceremony in Isle Luna. Then there was a machine that Exa was talking about. The one that had helped Theos fly.

Violet knew the names of the council members from the dinner conversations with her father. Their identities were not taught in school or to the general public, but they weren't a total secret. For security reasons, the identities of the council members were kept discreet and not blatantly announced. But their high positions were identified by their big golden apple pins.

Isle Luna had only one council member—Levida Luna. Violet remembered that her father always noted the rivalry between Levida Luna and Rosarius Santi, stemming from the controversial issue of religion. Levida promoted paganism and other religions, while Rosarius Santi wanted to bring the whole of Eden back to Christianity. Very controversial indeed, making a small faction of the council want to forego any religion for Eden.

If Levida Luna was indeed Theos's mother, then Theos, like Violet, would be one of the next members of the council.

Violet looked at Theos's message and sneered. She replied:

Yes, we do. Name your place and time.

Chapter 12

The Old Temple

At sa wakas / matatahimik ang mundo / Ano sa gayon ang kulay ng apoy / pagkapikit ng huling talukap? / Ano sa gayon ang tunog / ng hanging lumalakad / mag-isa / sa patay na bayan? / Ano ang lasa ng ulan / makalipas ang usok at kalawang? / Lukot na lansangan / liwanag na walang damdam / aliwalas / /

—Ralph Fonte, *Gunaw*

ZAP!

Violet found herself on the Isle of Templos. The sounds of teenagers partying almost deafened her. She was back to where it had all started. Where Theos had flown. Where they had dived. Where they had teleported midair. Where Exa had lost her contacts in the ocean. Where they had seen her green and violet eyes.

Before she could even look around, a warm hand suddenly grabbed her arm.

"Hey!" She started to resist but saw Theos.

"It's me," Theos simply said. Again, he had no shirt on and was clad only in white pants. He pulled her discreetly toward the temple.

"Why are we meeting here?"

"No one goes to the temple," Theos said. "Except for two teenagers who were just seen with the July twins before they were branded fugitives." He winked.

Violet fell silent. They made their way behind one of the columns of the crumbling temple.

"This looks stable," Theos noted, looking up at the column. Violet agreed.

Out of the way of prying eyes, Violet faced Theos to glare at him. She folded her arms across her chest. "Speak. I'm guessing big that you have something to do with this."

Theos sighed, throwing up his hands. "I suspect so," he said apologetically. "I never meant for this to happen. You have to believe me."

"You told your mom, didn't you?" Violet accused. "*The* Levida Luna?"

Theos sighed, looking upward guiltily. "I didn't mean—"

"Didn't you?" Violet prodded loudly.

The boy raked his braids with his hand. He sighed deeply. "Yes, I'm sorry!"

Violet exhaled. Well, it was the most logical way the council could have found out.

"But in my defense, I thought it was a gift from Eden," Theos said.

Violet's eyes narrowed in confusion and slight annoyance. How dense could this boy be?

"A gift?" she repeated. "From Eden?"

"Yes." Theos nodded. "Like my flying from Luna."

Violet rolled her eyes.

Theos glared at her. "Skeptic! You still don't believe me! Even after I showed you."

"Yes, you can fly." Violet rolled her eyes. "But Exa's eyes were like that since she was born. It is a mutation!"

"A gift," Theos corrected. "She has done nothing wrong—"

"And I agree!" Violet exclaimed. "But you know the rules!"

Theos looked at her blankly.

Violet scoffed, exasperated. "You're supposed to know, Theos! Is your mother not even trying to prepare you to be part of the council someday?"

Theos looked down. "She tries," he admitted. "But…uhm…I've been trying to avoid them." He looked at Violet. "Doesn't the council scare you?"

"So you have figured it out?" Violet's eyes narrowed. "About me?"

Theos smiled a little. "You're Violet Malaya," Theos said. "I knew it the first time Rufus introduced you. There are only a few Malayas in Eden. You must be Agham's daughter."

Violet nodded. "Well, yes."

Theos's eyes suddenly glinted. "And your father ordered an army of robots to the twins' house to capture them and then burn the house to the ground." There was an accusing tone.

Violet's head snapped up. "Hey!"

"Hey." Theos smiled. "I guess we both made mistakes trusting our parents."

"I didn't make a mistake," Violet shook her head. "How was I supposed to know that the security breach was them?

What could I have done?"

Theos shrugged. "Anyway, I still don't get why they're after the twins—"

"Because heterochromia is a mutation," Violet interjected. "And mutations are not allowed."

"An anomaly from Eden—"

"From Earth!" Violet corrected, exasperated. "Remember that even if Eden can edit our genes in the womb, our DNA map is still from our ancestors on Earth. Millions of years of evolution have also made way for certain mutations. That's why pregnant women do genetic screening, even lets AI edit, to correct mutated genes that would cause a form of disease later in life."

"She isn't sick," Theos noted.

"Yes." Violet nodded. "I think there's more to this than just heterochromia."

"Her parents?"

"Caught and brought for interrogation," Violet said.

Theos sighed. "So what do we do now?"

Violet looked him in the eye. "We learn as much as we can about the case. We also have to find a way to communicate with them."

Theos caught on. "They would have ditched the wristbands by now."

Violet nodded.

"I will try my best," Theos said. "Let's talk again some other time, shall we?"

Chapter 13

The Hang Glider

Then Moses led Israel from the Red Sea and they went into the Desert of Shur. For three days they traveled in the desert without finding water. When they came to Marah, they could not drink the water of Marah because it was bitter. That is why it was called Marah.

—Exodus 15:22–23

Three days had passed. Alec and Exa had been living on the edge of their island, trying to avoid possible cameras or any eyes. They had been scavenging for fruits to eat. They had never been this hungry before.

Most of the time, they tried to scour for new hiding places, hunting spots, or any means by which they could sneak into a store or something. However, their efforts were futile. Their island was more residential. There were fewer trees, vegetation, and plains in which to hide. Shops and markets were also few since it was convenient for people to teleport. Without a means to teleport, they couldn't leave the floating island except to jump from the edge and onto the waters below.

Once, Alec found himself gazing down into the depths. He had tried calculating the chances of making it alive if they jumped. He even thought of grappling down the island. Exa had to sigh and comfort him. "We can't. We have better chances of not jumping or grappling. If we reach the sea, where would we even swim to? Let's just hope they miss us."

Five times already, they had nearly encountered AI robots patrolling the island. They would be easily recognized for sure. The twins had to crawl among the tall grasses to escape as far as possible.

Sometimes at night, they talked about who could've ratted them out. Was it Violet? Or Theos? Most likely, it was the latter. Sometimes, they talked about their parents, praying for their safety. Had they made it out of their house? Did they escape? Were they captured? Sometimes, they just talked to themselves about what they had learned in school to distract their minds from their present situation.

By the third day, the twins knew every patch of grass on their island. That morning, they were eating freshly plucked berries and hiding behind a massive tree trunk.

Then they heard it.

"Detecting thermal presence," a robotic voice suddenly said from a distance. The twins peeked around the side of the big tree and looked at each other.

"The grasses are too short," Alec cursed. He looked around. The next tree was several yards away.

Exa looked up and pointed. "We climb."

Alec nodded. Exa took off her shoes and tied them to her arms. Then she began to climb, silently. Alec climbed after her. They found themselves among the tallest branches within minutes, looking down at a patrolling AI robot.

"We have to go even higher," Exa whispered. "It might look up. We'll have nowhere to hide."

Alec looked up at the higher branches. "We can't. Those are too thin and will break under our weight."

The AI robot started circling the tree.

"Why is it doing that?" Exa hurriedly whispered. "It knows! It knows!" She hurriedly moved to a higher branch.

"Exa!" Alec hissed. "You are going to fall, and we'll be doomed!"

"Then don't follow!" Exa hissed back. "One person's weight it is! Cover yourself with leaves!"

It took a long time. After the AI robot looked in all directions, it resigned itself to the possibility that it had detected a stray animal and went away.

The twins exhaled, not realizing they had held their breath for that long. After a few more minutes, Alec went down and looked around. The AI robot was nowhere to be seen. "Exa!" he called. "We have to move."

There was no reply.

Alec looked up. "Exa?"

Exa was so high up that he couldn't see her.

"Exa!" Alec called again.

Suddenly, there was rustling of leaves, and Alec saw his twin sister's foot. She was hurrying down with a big smile on her face. "Alec!" she exclaimed. "Look!" She pointed to the sky.

Alec looked up. The sun coincided with the big pink moon again, giving the whole expanse a blush. "What?"

"Someone's flying!" Exa pointed. "Over there!"

"Theos?" Alec looked at her like she was crazy. "Then we have to hide! He's probably— "

"No!" Exa exclaimed impatiently, still pointing. Alec looked.

Indeed, a small figure was gliding across the sky in the

distance.

Alec shrugged. "Hang gliding. Nice hobby."

They watched the hang glider land on a distant side of their island. "I think I know where he went," Alec said.

"Are you thinking what I am thinking?" Exa beamed.

Alec shrugged. "We're fugitives, anyway."

The two immediately ran toward where the hang glider had landed. They tried to stay undercover still, hiding behind trees and tall grasses.

Finally, they reached the spot. An orange hang glider was parked in the nearby meadow amid tall grasses and mimosa flowers that had mutated to continuously open and close their petals.

Alex gestured to her sister to remain behind him. Slowly, they started creeping toward the hang glider. A man was mumbling loudly, crouching down on the grass now and then.

Exa frowned. She met Alec's confused gaze. What was this man doing?

The man, wearing an orange suit, started plucking the mimosa flowers and putting them in a basket.

"…enough to get their extracts…" he was mumbling. The man stepped further and further from his hang glider.

The twins were nearly at the hang glider. Alec looked at Exa. He whispered, "We'll push the glider down the cliff and run together." Exa agreed.

The man seemed oblivious. He kept mumbling. "…that should do it. A little enhancer can probably mimic the effect…"

Alec stretched out his hand. It was trembling. Just a little closer, then they could leave this wretched island and hide somewhere safer.

Exa sighed and pushed her twin. Alec jumped a little. They could take hold of the handle.

Suddenly, a loud voice said, "I wouldn't do that if I were you."

The twins were startled to see that the man had turned around and was pointing a rifle at them. "STEP OUT."

Alec and Exa looked at each other and raised their hands. The absence of their wristbands was glaring.

"OH HOH!" the man exclaimed, not putting the rifle down. "What have we have here? Where did you two come from?"

Exa narrowed her eyes. How didn't he know them? Their names and faces were sent to everyone in Eden three days ago. Branded as outlaws. Wanted.

"We were just curious about your hang glider, sir," Alec lied smoothly.

"Why?" the man asked. "You're interested in hang gliding?"

"Yes!" Alec quickly nodded.

"When you could always…" The man's voice trailed off as he stared at their bare wrists. "Teleport."

"We…uhm…lost our wristbands," Exa began. "At the Sophronia Ocean. We just turned eighteen."

"Then how did you get here?" the man asked, smirking.

Exa fell silent.

"You're not good at lying, girlie," the man sneered. "Your twin has better chances."

The man's accent was quite different from anything in Eden. There were no other languages in Eden but one. The old world, Earth, they know from history, had millions of dialects and languages. They were taught that the presence of different languages could be a barrier to cooperation and success.

"How did you know we were twins?" Alec quietly asked. His facade was gone. "I thought you didn't know anything about us."

The man laughed for a bit, then roared. "Who doesn't? NOW STEP AWAY FROM THE GLIDER!" He cocked the gun.

Alec and Exa scurried away, their hands still raised. When they were three feet away from the glider, the man lowered his weapon.

The twins sighed and exchanged a glance. They lowered their hands as well.

"Wait," Exa said. "Aren't you going to report us?"

The man scoffed. "With what?" He raised his arms from his sleeves, revealing his also bare wrists.

The twins gasped.

"The July twins." The man smiled. "You two are famous where I came from."

"We are." Alec nodded. "Since three days ago."

The man roared with laughter. "That's funny! And why would you like to steal my glider? Hmmm? What island will you go to?"

The twins remained quiet. They hadn't really thought about it.

The man laughed again.

"Why are you making fun of us?" Alec asked, annoyed.

"Because you two are as pathetic and clueless as anyone in Eden," the man sneered.

"Why don't you have a wristband?" Exa asked.

"Because I'm like you!" The man smiled widely. "An outcast of Eden!"

The twins gasped. "There are more fugitives?"

"*Fugitive* is a strong word." The man chuckled. Then he looked around thoughtfully. "And we've done nothing wrong but to share this world."

He turned back to the twins. "Have you heard of Pangea?"

Chapter 14

The Interrogation

Then the Lord God said: See! The man has become like one of us, knowing good and evil! Now, what if he also reaches out his hand to take fruit from the tree of life, and eats of it and lives forever?

—Genesis 3:22

They were held not just for questioning but also for defying the law. They were criminals.

Agham Malaya looked at the hologram figures of Harold and Nia July held by red laser beams from AI-robot wrists on the walls of the bright white interrogation room.

It had been three days, but he still had no answer from them, to his frustration and anger. They refused to talk. Three genetic scans had been done since the couple was brought in. The result was not what Agham expected.

For days, AI had been scouring all the islands of Eden for unaccounted individuals or individuals without wristbands. But nothing. Agham sighed angrily, anticipating the threat. How hadn't Eden seen this coming?

There were burn marks on her wrists already. Nia's wristband, along with Harold's, was taken away when they were brought into the Tree of Life.

Undoubtedly, they were scanning their histories, messages, videos, and applications for clues. It was a good thing that they had learned to reset some settings quarterly, including deleting incriminatory evidence. And there were so many of them.

For the love of the twins. For the love of family. For the love of their land.

It was their motto. Her husband, Harold, was even more passionate and resilient in their duty. It was not even duty to them but a privilege. The twins were more than who they thought they were. Alec and Exa did not know what they symbolized.

The day the twins turned eighteen, despite Nia's claims that they would be fine, Harold had followed them to the *Voguer W*, keeping an eye on them from afar. The moment he saw Theos Luna, he knew there might be trouble ahead.

No, he didn't anticipate Exa losing her contacts in that dive. But knowing Theos Luna's mother was a high member of the Tree of Life Council, he immediately alerted Nia, and they took precautions, such as rinsing their house and wristbands of evidence. They had sent a warning to Pangea. They even got the passageway ready again, just in case they needed it. And they did.

Nia sighed, thinking of the twins. She had loved them for those eighteen years. Like they were her own. Where were they now? The twins were brilliant for their age, not as good as Violet Malaya, but they were more of risk-takers. They should be able to survive outside the grid of Eden's AI.

Nia's arms were aching from being chained to the wall. Every time her arms slacked off and fell a little bit, the red laser lines burned her skin.

She kept glancing at her husband, sometimes catching him

looking at her. But they had a strict rule about interrogation scenarios: Do not speak, as much as possible. Do not compromise the safety of the twins or of Pangea.

Nia leaned her head back against the cold wall, sighed heavily, and felt her strength wane. They weren't given food, just a little water occasionally.

Suddenly, the automatic door slid open, and the presence of Agham Malaya greeted them once more.

"AI robots have caught them!" he enthusiastically announced.

"NO!" Nia loudly gasped. Was this true? Were Alec and Exa caught?

Agham laughed like a maniac. No, he was probably bluffing.

Beside her, Harold started struggling. The red lasers began to burn his skin. "How could you!" he thundered. "They are just kids!"

"Kids who turned eighteen just five days ago." Agham shrugged. "Kids who should've been here in the first place."

"YOU ARE MONSTERS!" Nia screamed.

"No, you are." Agham looked at them straight in the eye. "You could've given these kids up to Eden. Our medical-grade AI could have saved them. In fact, Exa July should have been in the Intensive Neonatal Care on the day they were born—"

"YOU WOULD HAVE KILLED HER THERE!" Harold shouted in anger. "THE COUNCIL DABBLES IN INFANTICIDE!"

It came so quick that they didn't even notice until Harold's head suddenly reeled from the severity of Agham's slap. *BAM!*

"Watch yourself, July," Agham said coldly, deadly. "The

council doesn't take too kindly to baseless accusations."

"Baseless?" Nia spat out. "How many of those babies get to see the light of day again? Feel the warmth of their mothers' arms? How many, Agham? How many!"

Agham crossed the small distance from husband to wife and raised his fist.

"NO!" Harold screamed. "NIA!"

But Nia continued. "How many, Agham?" she cried. "If you check the statistics, almost none! Out of the many babies you put in there!"

Harold added, "Project Minus! The council's solution to population control!"

Agham roared. "I knew you were a spy, Harold July! You—"

"I work at the infirmaries!" Harold roared. "I saw everything with my own eyes! There was no secrecy! The rules allowed AI doctors to do the despicable—"

"To weed out those who are weak!" Agham roared. "They are screened for mutations and diseases–"

"As you know there will be!" Harold shouted.

"And stop them from multiplying! From mating with Eden humans!" Agham roared, his head up. Then his face suddenly looked psychotic and scary. "WE ARE THE SUPERIOR HUMANS, HAROLD! DON'T YOU SEE? WE DON'T NEED THESE KINDS OF MUTATIONS!"

"It isn't for disease," Harold sorrowfully reflected. "It is so you can build a superior human race."

"And what is wrong with that?" Agham reasoned. "THAT IS WHAT EDEN STANDS FOR! WE ARE PARADISE, HAROLD! WE ARE EDEN! WE ARE SUPERIOR TO EARTH!"

Agham sighed heavily, looking at the couple. He spat out, knowing his excellent sermon was lost on them.

"An Earth which was once filled with diseases and pandemics, hunger and poverty, wars and calamities." His voice was almost emotional. "This is the sacrifice our race has to undertake."

Nia and Harold still looked angry and resentful instead of inspired and motivated.

"Exa July and her twin," Agham said, "have no place in Eden."

"You are wrong!" Nia cried. "The twins are living evidence that they can exist in Eden! They're as good as anyone here! Smarter than most! Better than most! Despite her heterochromia, neither Exa nor Alec have never been sick!"

Harold added, "Don't you understand, Agham? You are flawed! There was no need to weed them out! Or anyone with mutation!"

"AI medicine should have helped those with mutations! To heal them!" Nia reasoned.

Agham looked at them sorrowfully. "You two are selfish! You are thinking of the minority. How about the good of all the millions living peacefully in Eden? You would tarnish their lives—their lineages—with disease! With poverty! With suffering!"

"That is not life!" Harold exclaimed, aghast.

"What would you know about life?" Agham shouted. "The two of you took on the roles of parents of mutated beings! Humans that shouldn't be allowed to live! And you know what's funny? YOU TWO WEREN'T EVEN THE ONES WHO MADE THEM! FOOLISHNESS!"

Harold and Nia fell quiet and quickly glanced at each

other.

Agham continued, "I wonder which pair you care for the most—the twins, or their parents." He looked at Harold. "You must be a blood relative—not the father but a blood relative. Several of your genes check out."

He looked at Nia. "But you! You didn't have much in common with the twins! You have sacrificed your life for them? Really, Nia? I saw who you were. You could've been a good botanist and a good researcher! We could have offered you employment at the Tree of Knowledge of Good and Evil! Your husband"—Agham looked back at Harold—"could've made rank in the medical field, but you chose to lie low. And, well, you have the genes of traitors!"

He sighed. "Not only are we still looking for the twins but for their parents as well. That case has always been a hard one to crack. But with the twins, we have enough reason to persevere in finding them."

"You will never get anything from us," Harold merely said.

"I know, I know," Agham said. He looked calmer. "You wouldn't sell out your own blood, Harold. And Nia wouldn't sell you out."

He began stroking his beard. "But you see, our scientists have developed a new technology. It's only in the early stages, though. If we could edit genes, why not…memories?"

"You wouldn't!" Nia gasped.

Agham looked at them and laughed like the villain that he was. "Clock's ticking, July. We will need volunteers soon. I will give you one day and one night to give me an answer. Where are the twins? Where are their parents? Where are Alexander and Miriam Leons?"

Chapter 15

Secrets

This day I call the heavens and the earth as witnesses against you that I have set before you life and death, blessings and curses. Now choose life, so that you and your children may live.

—Deuteronomy 30:19

Violet turned off the recording. She shook her head, processing what she just heard from her father.

Earlier that day, she had stuck a small recording bug on her father's cloak. Spying on a council member, much less the president of the Tree of Life Council, was tantamount to a crime deserving hefty punishment. Given her connections with Exa and Alec, she could even be labeled a fugitive.

But Violet was confident that she would never get caught. After all, who would dare? Her father would never expect it from her.

She sighed. Mr. and Mrs. July were not Exa and Alec's biological parents. Alexander and Miriam Leons were, but they, too, had vanished. Mr. July was biologically related to the twins, given the genetic similarities. It was not surprising that he would protect his kin.

Violet decided she would have to review the council cases to find out anything about Alexander and Miriam Leons. She would have to ask for Theos's help. Maybe. She didn't even know where to start looking for such information.

Next, she must visit the Tree of Knowledge of Good and Evil. She had to review the statistics for genetic editing, heterochromia, and perhaps even the records of Intensive Neonatal Care.

And what was Project Minus? Was the whole council aware of that? Was it really infanticide? She couldn't imagine such an atrocity. Would Rosarius Santi even agree to that, given his Christian values?

And also, the newest technology: memory editing. Her own mother, as one of the head scientists of the Tree of Knowledge of Good and Evil, should be aware of it.

Was Theos aware of this too? Violet wondered if her new accomplice would be able to gather anything.

Theos and his mother were incredibly close. He considered her his best friend. They had been together since the day he was born.

Lavida Luna was actually three hundred years old; her face, body, and health were maintained by AI medicine. She had given birth to Theos in her old age through reproductive AI. She'd had her eggs frozen for hundreds of years and fertilized in vitro. She had them genetically tested many times. Then, Lavida Luna spent most of her lifetime looking for the best candidate for a mate and sperm donor.

Theos never knew who his father was, except that he died right after giving his sperm to Theos's mother. That was what he was told. Truthfully, Theos never really cared, thinking it would be pointless.

He watched his mother raise him all by herself. He was her moon. Theos never had any reason to doubt or resent her.

That was why Theos felt incredibly confused and betrayed when Eden announced that Alec and Exa had become fugitives that day.

He talked with Violet first, then decided to confront his mother.

"Why would you do that?" he exclaimed when he saw her emerge from one of the Ficus trees.

Lavida looked blankly at her son. "Do what, Theos?"

"You told the council about Exa July!"

Lavida crossed the chamber gracefully and peacefully. "Theos, it was a matter of security. They had *long* been fugitives."

"Long been fugitives?" Theos exclaimed. "It wasn't their fault Exa was born with different eye colors—"

"No," Lavida interrupted. "It wasn't their parents' fault, either. Records have shown that they were genetically screened on a regular basis during the pregnancy. But it was their fault they're still here."

Theos looked at her in horror. "Do you mean to say—"

Lavida faced her son. "You are too young to understand all of these, Theos. This is a highly confidential council matter. How often have I advised you not to put your fingers in these businesses yet? Then when I'm trying to groom you for the council, the basics of being in the council, you go away." She threw a hand in the air.

"About that." Theos cleared his throat, knowing his mother had worked to give him the best normal childhood. Best, in that she had given him several advantages. "I met Violet Malaya."

"At the party, right?" Levida said. "Agham's daughter.

She's close to the twins. I wonder how she's dealing right now."

Theos felt quiet. For the first time, he felt a need to hide something from his mother. He could always tell his mother his latest antics, pranks, and escapades. In fact, his mother helped him escape justice for taking off the wristbands many times. But this, it felt…raw. Like it was a matter of life and death, not for him but for his new friends.

Levida continued, "But of course, Agham would be on top of things. He will not let anything happen to his precious daughter." Then she muttered under her breath, "If I were him, I'd use the Memoir Edit."

"The what?" Theos snapped.

Levida waved her hand. "Oh, it's a new technology we've been making. If we can edit genes, what would be the next best thing? Memories."

Theos looked at her in horror. "What for?"

Levida looked at her son incredulously. "Theos! Think of all the bad memories we can erase! Those things haunt, and they create trauma. We can free many people from trauma!"

Theos stared at her as if seeing someone for the first time. "You want to use it on Violet Malaya?"

Levida scoffed. "Not me. I'm not her mother. Her father, Agham, of course! It must be dreadful, looking back and realizing you've been rubbing elbows with fugitives."

Theos took a deep breath and stepped backward. "I brought Exa here to witness the moon ceremony. Are you saying you will put me under that memory tech too?"

Levida looked at him. "Well, don't you want to? We can try! It's still in phase 2, though."

"You will put me under a phase 2 technology?" Theos

snapped incredulously.

"It passed phase 1." Levida rolled her eyes. "It's safe, Theos." She walked to him and softly cradled his face in her hands. "I will never put my son, my only son, at risk or in any danger."

"Then—"

"But Exa July is a danger." Levida's eyes narrowed.

"It's not her fault!" Theos pushed his mother away. Levida looked surprised.

"It could be a blessing, just like flying, which came from Luna," Theos reasoned.

Levida suddenly smiled bitterly at him. "Believe me, some gifts are curses. And some gifts, well, they're earned."

Chapter 16

Pangea

In the whole of geophysics, there is probably hardly another law of such clarity and reliability as this—that there are two preferential levels for the world's surface which occur in alternation side by side and are represented by the continents and the ocean floors, respectively. It is therefore very surprising that scarcely anyone has tried to explain this law.

—Alfred L. Wegener,
The Origin of Continents and Oceans

"AHHHH!" Alec and Exa screamed as their feet took off from the ground.

In between them, the man who owned the hang glider roared in laughter. "This will take a few minutes!"

The three of them clutched the hang glider rods as it soared through the sky. Alec and Exa watched as their floating island grew smaller and smaller.

"Where are we going?" Exa screamed into the air as she watched the vast sea below.

"I told you! PANGEA!" the man shouted as he tilted the hang glider. "We're near!"

"Where?" Alec shouted. He could see a desert coming up. But it looked empty.

"There!" The man pointed.

Exa glanced at her brother. Were they right to trust this man? But it was this or being stuck in the island, being hunted by robots. The hang glider lowered until they were in the middle of the desert. Then, slowly, they landed. It was hot and arid. The hottest place the twins had been to. The man noticed their discomfort.

"One of the places Eden has overlooked." He smiled sadly. "AI can never do it all."

The twins looked at him apprehensively.

"Where have you taken us?" Alec asked, looking around.

"Don't worry!" the man said. "AI robots rarely go here. Almost no one goes here. We're at Casiah Desert. I'm sure you have heard of it."

Casiah Desert was an unlivable desert in Eden. It hadn't been developed by AI. But it was still being monitored from time to time.

"I thought you said Pangea," Exa said.

"I did!" The man grinned. "Look here!"

He walked over to a certain spot, and suddenly, a white space opened.

The twins blinked. "What—how?"

The man grinned. "It's a mirror room." He went to a wall and knocked on the mirror.

The twins blinked, suddenly realizing they were in front of a mirror building with almost no edges seen.

"Let's go inside," the man said suddenly. "Quickly! It's safer here."

The twins followed him into the white room, which had a staircase going down the middle.

"Pangea is underground," the man explained. "We built this place to hide from the council."

"Underground?" Exa repeated.

"Yes." The man beamed. "Have you guys been underground?"

"We've heard of the underground in the floating islands," Alec said. "The area should be small."

The man shook his head. "That is for a floating island. We're on the surface of Eden. The very surface." He began to descend the staircase.

The twins followed him. It was a long staircase that reached another mirror at the end.

The man suddenly pressed his palm on the right corner. The mirror slid open, revealing…*a new world beneath.*

"The underground is vast!" The man beamed. He turned around to the twins. "By the way, my name is Liam. Welcome to Pangea!"

Chapter 17

The World Beneath Their Feet

Una na nating inibig ang abot-tanaw
bago pa man sukatin ang daigdig
ng ating mga katawan.
Ang bawat mapa ay isang mundong iniukit
sa balat pagkatapos ng pagtataka: bukal
ng ilog at ng amihan, kakatwang hugis
ng pulo sa alapaap, anino ng malalayong kulog

—Ralph Fonte, *Kartograpiya*

From below, one could not see the familiar pink skies and pink moon of Eden. But millions of tiny lanterns were above them, so light wasn't an issue.

The twins expected a small, cramped space but not a city made of earth. If Eden had floating islands, Pangea had many houses built on earth, carved and molded. Structures all sharing the same crust, same soil. An underground alive! A different world just beneath their feet, all along!

"We also placed a little concrete there, and then some glass," Liam explained, showing them around, pointing to the houses. "But earth is earth. Rock is rock. We haven't had any infrastructural accidents since, I think."

Several people paused to look at them. Some children

were running from caves and houses and looking out of their windows at them. The best thing was that no one had any wristbands here.

Exa bowed her head, suddenly conscious. She and Alec held hands.

"We're a different community here in Pangea. We're a combination of the old Earth and Eden. We're small, but we're tight. We get things done, and we survive," Liam said.

"How many live here?" Alec asked. He was almost surprised his voice didn't echo. He and his twin could breathe easily. It wasn't cramped or damp. The air around had the right amount of humidity.

"The latest population count was three hundred sixty-one," Liam answered. "You have to understand there are not many fugitives in Eden. Although the Pangea community has been forming for over three centuries, people here have the common sense to know that population control is necessary. We are still fugitives, in a sense. Undocumented. Unnamed. Unaccounted for. You wouldn't want to start a family here if you have the options of AI medicine, AI education, and the like. Most of us keep up to two kids only."

"Who runs this place?" Alec asked.

"The elders," Liam answered. "They have been waiting for you."

"Why?" Exa asked. "You mentioned something like that a while ago."

"The elders should be the ones to explain it to you," Liam said.

"They're like the Tree of Life Council?" Alec said.

"No." Liam shook his head. "The elders inform us, the

people. When it comes to the well-being of Pangea, we have a vote. Not only the elders. Once a week, we all gather in the town hall and discuss some issues."

"We will meet the elders today?" Exa asked.

"Of course," Liam said. "I'm taking you to them. Where do you think we're going?"

Alec and Exa looked at each other. Their concerns were mirrored in each other's eyes. They were far away from what they knew as Eden, and there were no AI robots to chase them here. But danger might still be lurking.

After passing several houses, they found different tunnels.

"Each tunnel leads to a specific place," Liam said. "They are actually numbered." He pointed to the numbers scrawled on top of them. "One means infirmary. Everything health-related is in there. We don't have AI medicine or genetic screening." He looked almost pitifully at the twins. "But we can heal."

"With what?" Alec asked, raising an eyebrow.

"Antidotes and antibiotics," Liam answered. "The Earth way. We try to gather plants, venoms, and the like. Then we turn them into medicines. The pharmacy where I work is inside that tunnel too. I have made it even bigger. I've been tinkering with different potions all my life. That's why you saw me picking mimosa flowers. They're good for memory loss, but I have to find something that can make its extract more potent for the nervous system and, of course, safer."

"You've been a quack doctor," Exa noted.

Liam sighed, stopped, and turned to them. "I hate it when you people say that."

The twins glanced at the man's rifle to make sure it wasn't

pointed at them.

Liam continued, "One cannot deny that AI technology is much faster and more efficient. Imagine curing cancer in a day. Dang, you kids don't even know cancer! Diabetes! Heart failure, or kidney disease! But it doesn't make our medicine less effective. Ours take time, but as you can see, our medicines have fewer side effects. Fewer mutations." He looked straight at Exa.

He fell silent momentarily, then said, "Ever wondered how, in a perfect paradise, Eden maintains its population?"

"It doesn't," Alec replied. "There are no rules regarding reproduction—"

"And abortion is normal." Exa shrugged.

"Yes! You've been taught the choice of abortion. Oh, but not only abortion!" Liam exclaimed. "Why do you think they want *your* heads? No mutations are allowed in Eden! No one less than perfect is allowed to live!" He looked again at Exa again.

"How do you know about us?" Exa whispered. She felt a little conscious, as if Liam could see her heterochromia right through her lenses. She clutched her brother's arm.

"Who doesn't?" Liam spat. He started walking again.

"This tunnel"—he pointed to the second tunnel—"is the engineering section. If you want anything to be made, approach them. That includes tech. We don't really use AI tech because it can be traced. Signals are tricky here as well. But we use the remnants or simple technology to help around here. We're not entirely decrepit."

"Are you sure?" Alec said, looking around. Pangea, with the absence of the technology they were used to, felt ancient. But somehow, it still had a sense of strength and sturdiness to it, like how this city stood and existed unapologetically in contrast to Eden.

"Hey, watch it!" Liam pointed at Alec. "Everything here is manmade! Not AI! Have a little respect!"

"Oh, I do," Alec said quietly, continuing to admire the earthly walls and carvings.

Up above, they found teenagers on a wooden platform going down. "See?" Liam said. "That's our elevator here."

Alec and Exa nodded.

Liam smiled. "This isn't stone age. Just…Earth age, but underground, and more advanced, but not AI."

"We get the gist," Exa said quietly, nodding.

They continued walking. Liam pointed to the third tunnel. "That leads to the farmlands. It's a beautiful place. You might want to explore it in the next few days."

"How do you farm here?" Alec asked, genuinely curious.

"Not with AI robots." Liam rolled his eyes. "We built an intricate irrigation system and engineered a system to let the sun shine in once in a while. It's beautiful there when it rains."

They reached the fourth tunnel. "This one is like the mall, like a marketplace, or a trading area." Then he pointed to the fifth one. "This one here is the school and library."

"An actual place?" Exa asked, thinking about their holographic school.

"Yes," Liam said impatiently. "An actual place. Where kids sit on actual chairs. There's an actual board. And actual paper. There are actual books. Not holograms."

"How old are your books?" Alec asked.

Liam smiled. "You'd be surprised. Most were stolen from

the Tree of Knowledge of Good and Evil." Then he laughed. "The council forgot that books, not just AI, are good sources of information."

He pointed to the sixth one. "Here is where we bathe. There's an underground stream inside. It becomes a river if you follow it. We usually close this tunnel during rainstorms for the risk of flooding. But for centuries, no flood was ever recorded."

They turned to the seventh one. Liam paused and gestured to the twins. "Come, I believe the elders are waiting."

"How? How were you able to inform them?" Alec asked.

Liam took a tiny square thing from his pocket. "We have phones, or like pagers, courtesy of our engineering department. Besides, word travels fast here."

"Phones," Exa repeated softly. "Pagers." Words that felt foreign on her tongue.

The three of them entered the seventh tunnel into a large clear opening.

The twins found themselves gaping and looking around and above them.

They had entered a sort of arena surrounded by bleachers made of earth, like a stadium, going up and up. There was a glass ceiling above, allowing the sun to pass through and illuminate them.

Several people were scattered on the bleachers talking among themselves. They fell silent when the twins walked in. There were at least a hundred of them.

Liam looked up at them all. He said in a loud voice, "I return with my mission accomplished."

The people looked expectantly at him.

"I have gathered a basketful of mimosa flowers!" Liam said enthusiastically.

Several onlookers groaned. "Liam! Stop kidding around!"

Exa and Alec smiled and sniggered. But Alec looked at Exa in worry. A mission. These people knew them, and this man had managed to get them to go with him. They had become the prey.

"What?" Liam feigned innocence. "Mimosa flowers are no laughing matter!"

Several people groaned.

"All right!" Liam announced. "I have brought the twins!"

An old man stood up from the third-row bleachers. He said in a booming voice, "WELCOME TO PANGEA, ALEC AND EXA JULY! WE HAVE BEEN WAITING FOR YOU!"

Alec and Exa took him in. He looked old, as if he had seen hundreds of years. But from Liam's explanation of the lack of AI medicine, he was probably just decades old.

"How do you know us?" Exa repeated her question from a while ago.

"We know your parents." Another woman stood up from the same bleacher. "I am Elder Nora, and this is our Highest Elder, Ajax."

Alec and Exa gasped, looking at each other.

"Our parents?" Exa repeated. "You know them." All her worries came back to her.

"We have been following you. Your parents have been watching you all your life—"

Alec frowned. "Well, of course, they have—"

"Are they still alive?" Exa interrupted, almost shouting.

"How do you know they're safe? Are they here? They weren't taken by the robots?"

There were several murmurs.

Another man stood up. "I think they're thinking of Harold and Nia."

"Who else!" Alec said.

Highest Elder Ajax sighed. "Your uncle and aunt have fiercely fought and protected you for eighteen years. We were deeply saddened when we learned of their capture."

The twins felt silent as they tried to digest what they'd heard. And oh, there was the echo Alec had been waiting for since they had gone down. The echo in the silence. In disbelief.

"What do you mean aunt and uncle?" Exa suddenly demanded. "They are our parents!"

The elders started shaking their heads.

"Your parents sought refuge here after they gave you into the keeping of your uncle and aunt," Ajax explained. "You don't even remember what they looked like or their names. But your family is of Pangea. Your parents became one of us. Your aunt and uncle are one of us."

"One of Pangea?" Alec repeated.

"Yes," Ajax answered.

"Our parents are here?" Exa clarified. "Our real biological ones?"

The elders looked at each other. "They went on a mission four days ago. Your uncle warned us that you had been seen by a member of the Tree of Life Council."

"So where are they now?" Alec asked.

Ajax sighed. "They were supposed to return yesterday, or

with you. But we haven't heard from them. That's why we sent Liam to bring you back and to try and locate them."

Liam suddenly spoke. "I found Miriam's tracker at their old burnt house. But that's the last lead."

Alec and Exa couldn't believe their ears. Their real mother was just at their house?

Real mother? As if the last eighteen years with their supposed mother and father weren't real. Exa wanted to shout. In the last three days, they were attacked, lost a home, lost their parents, saw an unbelievable underground civilization, and then got told their *real* parents were here. It was too much to take in.

Ajax saw their discomfort and confusion. He reassured them, "We will try to find them."

"It's only been a day," Nora said. "No news is good news."

Ajax continued, "As for Harold and Nia, we have been planning a rescue mission for them."

At this, Alec and Exa looked up in surprise. "You have?" Alec repeated. There was hope in his voice.

"We will discuss the plans with you later if you want," Ajax said. "But for now, we have finally met you, Alec and Exa."

The elders smiled and greeted them. "Welcome home!"

However, the twins were unable to return the enthusiasm. Home was a sacred place they had lost days ago. Home was Harold and Nia July, now taken somewhere out of their reach, with unspoken horrors befalling them. Home was, they just learnt, an adoptive uncle and aunt, because their own parents had to hide. Home, despite all Pangea could offer, still felt a little out of reach for them.

Chapter 18

Eighteen Years Ago and Now

The word "freedom" is forbidden to salmon, along with words like resistance, escape, disobedience, defiance, destruction, revolution. If they used these words, Big-Mouth warned them, not a single one of them would be left alive to return to their home river and spawn.

—Ahn Do-hyun,
The Salmon Who Dared to Leap Higher

Alec and Exa didn't ask any more questions at the town hall. The six elders introduced themselves and smiled at them. Then they asked if they had eaten anything that day. When the twins shook their heads, another woman led them to the banquet hall inside the marketplace tunnel.

The twins were surprised that Pangea had abundant food, mostly vegetables. After they had eaten their fill, Liam led them to their own small house sculpted from the earth. It was a tiny one out of many.

Some people, their neighbors, looked and smiled politely at them. Their tiny house had a clean red rug at the center and a small bedroom with two simple mattresses. It had a small bathroom and toilet with an efficient water system.

"I'm sure this is nothing compared to what you are accustomed to," Liam apologized.

Alec and Exa reassured him that this was all right, compared to lying on the grass and keeping one eye open for AI robots.

"I'm sure that was a lot to take in," Liam said, empathetic for the first time. "We will give you time to adjust. You can explore Pangea. When you're ready, you can talk to the elders." Then he left.

Alec and Exa looked at each other. Exa wanted to cry, but for now, they were saved.

"It's okay," Alec said, trying to comfort her. "We'll get to the bottom of this."

"Did you have any idea?" Exa asked. "Any inkling at all?"

Alec shook his head. "None. They were so good to us. They had given us everything."

Exa nodded in agreement. "I never felt we weren't theirs."

Alec sighed. "Let's get some rest. Maybe tomorrow will have more answers for us."

Exa felt exhaustion overriding her body. The days of hiding and running had taken a toll on her. She nodded as she lay down on the mattress. A bath would be good, but they would have to explore that tunnel later.

However, Alec wasn't able to get a wink of sleep. He lay on his mattress, wondering what the future held and what other secrets there were.

After a few hours, they heard voices outside, passing their house.

Exa yawned, sitting up. "What's happening?"

At that moment, a soft knock sounded. Alec stood up and opened the door. It was the shy, kind woman who had escorted them to their meals a while ago.

"Hi, I forgot to introduce myself earlier. I'm Annisa. The elders asked me if you would like to join us for dinner. We usually gather at the dinner tunnel," she said and smiled. Then she quickly added, "But if you want some time alone, we could—"

Alec raised a hand. "No, it's okay, we'll be there."

Annisa smiled at him and Exa. "Okay, we will see you there." Then she walked away.

"Are you up for dinner?" Alec turned to Exa.

"Let's get it done," Exa replied. "We have to learn about them and this place."

"We have to know more about our real parents," Alec said.

"And learn about the rescue plan for *Mom and Dad*," Exa said.

The twins headed to the dinner tunnel, where it looked like a feast had been served. Several heads turned when they appeared. Nevertheless, everyone continued chatting and eating.

Liam saw them and raised a hand to invite them to sit beside him. The twins went over.

"…just a little bit of the extract!" Liam was saying as the twins sat beside him. He absentmindedly handed them plates and utensils. He was talking animatedly with the elders. "It's worth a try! We can sneak into one of the labs, get the gene modifier, enhance the gene, and test the potion!"

"It would certainly be a good prospect," Nora replied. "Especially with the reports we got last week."

"The Tree of Knowledge of Good and Evil has reached phase 2 trial of their memory-changing technology. The Memoir Edit," Ajax said.

Liam nodded. "This new potion involving mimosa flowers can help counteract, maybe." He shrugged.

"The Tree of Life Council likes changing things, editing things!" one of the elders, Yvan, a big man, complained. "Putting their hands where they don't belong!"

"I wonder how they would test phase 2," another man said in a small voice as he scooped up some beans. He was tall and lanky, with thick glasses (something the twins very rarely saw in Eden).

"What do you mean? They have all sorts of lab animals," Ivan grunted. He reached out for another chicken leg.

"Well, how would they test the memories of an animal?" the man said to no one in particular. "Hmm, maybe they'll see if a mouse can remember the path in a maze to a piece of cheese? But there are all kinds of memories. Short-term, long-term, muscle memory. It doesn't seem…quite…hmm, efficient? To use an animal."

"Are you saying they'll test it on humans?" Ajax suddenly asked.

"It's a possibility." The lanky man nodded. "It's the best course. The Council already has…uhm…questionable ethics."

"Who? Who will they test it on?" Liam asked.

The twins suddenly realized, as they exchanged a worried glance, that the test subjects would be prisoners—their mom and dad, or rather, their foster parents.

Ajax and Nora noticed their discomfort.

"Thank you, William," Ajax said. He turned to the twins.

"We have already laid out the rescue plan for Harold and Nia, Alec and Exa. We will run it by you after dinner if that will ease your worries. Meet me at my house."

Exa suddenly spoke. "We want to come along for the mission."

Liam snorted. "You and your brother?"

Alec glared at him.

"They couldn't even steal my hang glider," Liam told the elders.

"It should have been easy, but you had a rifle!" Alec reasoned.

Liam looked at them. "Precisely, kid. Those guys out there mean business. You're lucky I only had a rifle. What about the lasers they attacked your home with? An AI robot army? A bomb? A grenade?"

Nora nodded. "The twins need training if they want to be out there. But personally, I'd feel better if you stayed here."

Yvan addressed them. "Your parents made a lot of sacrifices for you. We don't want that to go in vain."

"We've never seen our parents—" Alec started.

"Doesn't mean they weren't present," Ajax finished. He looked at them sternly. "Your parents took on jobs for Pangea, which meant visiting Eden to see or visit you. Why do you think you have an escape tunnel underneath your house?"

"They were supposed to pick you up at the end of the seventh tunnel three days ago," Nora said. "But we think something more worrisome is up."

"Like having your foster parents imprisoned," Ajax said. "The Council can use many devices on them just to get them to spill secrets."

"What did our parents do?" Alec suddenly asked. "Why are they being hunted? How did we end up with our aunt and uncle?"

The elders looked at each other and sighed.

"Nia and Harold should have done this talk days ago," Yvan complained as he looked back at his plate.

Ajax sighed. "Eighteen years ago," he began, "January 31, 722 AE, your mother gave birth to you. Your parents were average Eden citizens who relied on AI for everything. They chose your hair color, blond after your mother, and dark blue eyes.

"Your uncle, Harold, was a medical assistant. He had AI working for him. You realize that although most medical procedures, like your birth, can be done by AI, they still need to be supervised by humans—not all procedures, just major ones. Even Eden knows they cannot leave AI alone.

"Your uncle was a bit rebellious, wayward, in their family, especially during his teenage years—"

"I can't imagine," Exa muttered to her twin.

"Oh, he organized wild parties at the Isle of Templos and at the *Voguer W*!" Nora exclaimed.

Alec and Exa exchanged wide-eyed glances.

"He's quite the adventurer!" Yvan confirmed.

The twins couldn't believe it. Was their foster father, always a bit of a worrywart, an adventurer?

"He's my best friend," Liam added silently.

The twins immediately turned to him.

Liam smiled sadly. "We met when he turned eighteen

years old. Annoying fella he was back then. He found out about Pangea through me. Despite his misadventures, he got into the medical field but never pursued higher positions."

"The existence of Pangea was forbidden," Ajax explained. "And kept secret. Rumors of our existence were controlled by the Council."

"At first your foster father, Harold, tried to tell his sister, your mother, Miriam, about Pangea," Liam said. "She never believed Harold. She only thought he was up to no good, as always. Until the day of your birth."

"Your parents did everything right prenatal. Consultations, genetic screening, and yet you were born with heterochromia," Ajax said.

Exa wanted to touch her eyes. She had never taken her contacts off since the day they left home, so she was lucky not to have an eye infection by now.

Ajax continued, "They were suggesting Intensive Neonatal Care for you. Alec was free to go, but since you two are twins, they would still keep an eye on him."

"The Intensive Neonatal Care," Liam began, "oddly, doesn't have any human medical doctors or personnel. It was run by AI and the Council."

"Odd because you would expect serious medical maladies from neonates or babies whose immune systems are still too weak," Nora said. "A job that needs human care."

"Harold was suspicious when he first found out about it. Then he gathered the data from medical centers. Only about 0.01 percent of babies who go in there came back alive and well to their parents," Liam explained.

"He did that?" Exa whispered.

"He revealed the statistics. Sent it to us. He couldn't announce it to Eden. But rumors have started," Ajax said.

"The Council has always been careful not to leak Project Minus," Nora said.

"Project Minus?" Alec repeated.

"A project of the Tree of Life Council dating back hundreds of years, confirmed by our source in the Tree of Knowledge of Good and Evil. A project that eliminates those born with mutations," Ajax explained.

"Infanticide," William groaned and shook his head. Exa and Alec turned to him. It was the first time they had heard him speak since his concern about the Memoir Edit subjects.

"You would have been killed." Liam looked at Exa in the eye. He turned to Alec. "And you would have been killed too, sooner or later."

The twins fell silent.

Liam continued, "When Miriam found out where they were taking you, knowing what she knew from Harold's findings, she wouldn't let you go. Your real father, Alexander, smashed the AI doctor's head, and then your parents tried to escape with you. But that was almost impossible with AI cameras and detectors everywhere.

"Your mother gave you to your uncle. Your uncle and Aunt Nia hid you. It was Harold who led your parents to Pangea to escape. Since then, the Council has been looking for you and your parents.

"Harold listed you months later in the system, saying that Nia gave birth at home to twins. Since it was a home birth, no AI was present to double-check the facts. Luckily, Nia was also pregnant at that time, but she had lost her own child. Her prenatal records check out that she was with a child. So it wasn't surprising she supposedly gave birth later," Liam said.

"Mom lost a baby?" Alec whispered.

The elders nodded.

"I treated her," William said. "With Liam. Harold brought her here when she bled."

"Poor Nia," Liam sighed. "She then treated you two as her own."

"Our real parents?"

"Checked on you at least once a week." Ajax smiled. "It was a sacrifice they took so you two can have a life above the ground."

"We have to save them," Exa suddenly said, thinking of Harold and Nia.

"We know," Nora said gently. "We told you we have a plan—"

"And we want in," Alec insisted.

The elders looked at each other. Liam groaned.

"I oppose," he said. "The twins might sabotage the plan. They know nothing!"

Suddenly, Ajax had an idea. "Liam, have you talked to the engineering department? I heard Heph is tinkering with new technology. The twins might be the perfect for that job."

Chapter 19

Heph and Hypatia

Ill deeds thrive not. The slow catches the swift; even as now Hephaestus, slow though he is, has outstripped Ares for all that he is the swiftest of the gods who hold Olympus. Lame though he is, he has caught him by craft, wherefore Ares owes the fine of the adulterer.

—Homer, *The Odyssey*

After dinner, Ajax and Liam escorted the twins to the engineering tunnel.

"I have doubts about this, Ajax," Nora had admitted before they left.

The elders voted right there and then on whether Alec and Exa could join the rescue team. Only two voted no. The rest strongly advised that the twins be heavily supervised at all times and that if an imminent threat was apparent, a member of the team should immediately escort them back to Pangea.

"I don't understand," Exa had said, exasperated. "We want to help! Those people raised us!"

Ajax had smiled kindly at her. "We understand your impatience. But for the sake of Alexander and Miriam, we must keep you safe. You two are also Pangea's greatest evidence! That those with mutations may be allowed to live."

Alec and Exa were startled. They had never considered it that way. Ajax made it seem like their presence and existence was tantamount to Pangea's revolution against Eden.

Heph was a big man whose face was almost covered with beard and grime. His hands were dirtied and calloused. His clothes were dirty with burns and black stains. His eyebrows were knitted together. Focusing on his latest project, he continued tinkering at the table.

The twins were sure he hadn't been at dinner. He was unmistakable.

Ajax cleared his throat. "Heph," he began. "I have found you two volunteers."

That was when the man raised his face to stare at the twins. "Why them?"

"They are Alec and Exa July—"

"I know who they are," Heph grunted. "Patia had already warned us about them and the Council."

"They wanted to join the mission," Ajax explained. "They are also your best bets, as they have used Eden's wristbands beforehand."

Heph gave a loud sigh as he moved his almost gigantic hands, putting his latest project on full view.

"Wristbands?" Alec said incredulously. He stared at the black bands on the table that looked like theirs back in Eden.

"Even better than Eden's," Heph grunted. "I was able to hack their system, thanks to Patia. But these bands can escape any AI detection. You can be on Eden's system, but they'll never see any new users."

"Isn't it amazing?" Ajax beamed.

"This is a game changer," Liam whistled. "We can have a better playing field."

Heph nodded. "We can try." Heph cut a red wire on each of the two wristbands. "That should do the trick."

He gave Liam, Alec, and Exa each a wristband.

Liam looked at it warily.

Alec and Exa looked at each other and smiled.

"Go on, put it on!" Heph prodded. "What are you waiting for?"

The three put the bands on. The wristband blinked red as soon as it touched their skin, indicating life.

A purple holographic screen appeared before them. Words took form.

Set up identity…

"Choose a code name to give it," Heph instructed. "That way, it cannot be traced to you."

Liam looked at the twins. "Perhaps they should also do disguises."

Ajax nodded. "Agree. Annisa can help you with that at the other tunnel."

"Artemis," Exa said.

Alec raised an eyebrow. Exa grinned at him. It had always been their inside joke.

"Apollo," Alec said, rolling his eyes.

"Harry," Liam said softly. The twins glanced at him. Only their foster mother had ever used their father's nickname.

"Now you're free to check it out," Heph said, interrupting the moment. "Eden can never detect you."

Alec went to the travel options. "Cool! We can teleport!"

"Yep." Heph nodded. "But it only allows for the teleport points of Eden. Or if you set up a certain exact point. Nevertheless, it should be located in the AI grid of Eden."

The twins nodded. "We can message each other?" Exa asked.

"Yes, use the code names," Heph answered.

Liam did his scrolling. "All right, almost everything is in here."

"Everything is in there," Heph corrected. "And they can never catch you."

Ajax beamed. "Please extend our gratitude to your sister."

Heph nodded happily. "Sure."

"Hypatia is our source in the Tree of Knowledge of Good and Evil. She lives as a citizen of Eden, but she is also one of us," Liam explained.

"Separated?" Alec asked.

"No," Heph answered. "She infiltrated Eden when we turned eighteen."

"I've known Patia since she was a kid." Ajax smiled. "She was goal-driven. A dreamer. An achiever. Just like Heph here."

Exa raised an eyebrow. They talked about this woman as if infiltrating Eden was a career. Well, was it?

"She is still the only one to have done so." Heph smiled. "We had different dreams. I stayed here in Pangea. I am much more needed here."

"And it is better to have an outside eye," Liam said, nodding.

"How did she do it?" Alec asked.

"Perhaps a story for another day," Heph said, still smiling. "I am just glad those wristbands work for you three." He turned to Ajax. "I will try to produce as much as I can. A wristband can take a month to make and reprogram."

Ajax bowed his head at him. Alec found himself staring at the Highest Elder bowing to another man. "Thank you." Ajax smiled warmly.

He turned to the twins. "Now, for the rest of the plan."

Time is of the essence, Violet realized as she mulled over the interrogation over and over. Her father was not a patient man. Mr. and Mrs. July would stop at nothing to protect the twins. Their eighteen years of taking care of them was proof of their dedication.

This was a dangerous mission she would undertake. Honestly, looking at her laid-out plans, anything could go wrong. But if she didn't hurry, Mr. and Mrs. July would be guinea pigs.

That night, she found her parents in discussion in the library.

"—already passed phase 1 trials already," her father said between gritted teeth. "You know there are no better substitutes for memory subject—"

Her mother had a hand on her father's arm. "No, it is inhumane. We can try other methods. Human beings, despite being the most superior, are not the only ones with an extensive memory—"

"These are prisoners, Azula."

Violet had guessed enough. She cleared her throat.

Her parents looked up and saw her.

"Violet, shouldn't you be in bed?" her mother said, kindly.

Violet took a deep breath. "I was thinking, Mom and Dad, about what is happening. My best friends turned out to be fugitives…"

Azula threw an I-told-you-so glance at her husband and sighed. "Listen, we're sorry—"

"No, it's okay, it's okay." Violet held out her hands. "I was thinking how irresponsible it was to associate myself with them. And we know, as your daughter"—she paused and looked Agham in the eye—"how important our roles are to Eden. Eden will always be of the highest priority."

Agham smiled proudly at his daughter. "What are you suggesting, Violet?"

"I want to begin my training now," Violet announced. "As a future member of the Tree of Life Council."

"You're not yet eighteen." Azula narrowed her eyes. She turned to her husband. "I thought we agreed that she would begin training after she was eighteen and had passed all her academics."

Agham couldn't be prouder of his daughter than at that moment. He turned to his wife. "Your daughter is the one requesting this, not me." He grinned widely as if he had won the lottery.

"I want to start tomorrow," Violet insisted. "I am also worried about the two of you. The sooner I can learn, the sooner and the better I can help serve our planet, Eden."

Agham nodded. "You have definitely grown, my daughter! No wonder you are the best. I agree with your decision. The first step would be to review the history and laws of Eden."

"I know those already," Violet found herself groaning.

Agham chuckled. "Not everything. I will give you permission to use the archives of the Tree of Life Council tomorrow."

"And also the Tree of Knowledge of Good and Evil!" Azula suddenly said. "If we're going down this path, I suggest you start to spend time also in the Tree of Knowledge of Good and Evil. I will show you around so you can learn. Absorb everything, Violet. After all, articifical intelligence runs Eden." There was a slight change in her expression, but Violet could not put her finger on it.

Violet smiled instead. "That would be lovely, Mom. I can go with you in the mornings."

Azula beamed at her. "Very good, Violet."

She was supposed to learn all of these things in another year. But given the situation, Violet knew she needed more information. An inside knowledge.

The next morning, Azula and Violet teleported themselves in front of the Tree of Knowledge of Good and Evil, which was just a few meters away from the Tree of Life Council.

When she entered, Violet gasped as she finally saw the Tree of Knowledge of Good and Evil for the first time. Almost everything was made of glass or holograms. Instead of box-type rooms, every room or section was spherical and connected to the adjacent spheres.

Azula led her daughter to every sector and started explaining their new projects, the most notable being the latest Memoir Edit.

"The Memoir Edit has passed the phase 1 trials. We used

seven thousand rats for that phase. It's now ready for phase 2," Azula said.

"Are rats still going to be your population for phase 2 trials?" Violet asked. "Is it a bit harder to test Memoir Edit in those who do not have an extensive vocabulary and those who cannot distinguish between long-term and short-term memory?"

Azula frowned and agreed. "Your father and I are aware of those limitations. But let's see." She looked at her wristband where a tiny holographic map showed her a red dot getting closer.

Azula continued, "Subjects to be tested on were promised by twelve noon."

Twelve noon. Mr. and Mrs. Exa July would be in transit an hour before. Violet smiled to herself. She had timed these things really well.

Chapter 20

The Rescue Mission

Who wishes to fight must first count the cost.

—Sun Tzu, *The Art of War*

ZAP!

The wind opened and threw Alec, Exa, and Liam out of thin air. The twins smiled; they had missed teleportation, the rush, the thrill and the sheer efficiency of it. Liam groaned, looking close to puking his guts out. He wasn't used to it—his molecules shrinking and being made to pass through the space continuum.

ZAP!

Since Heph was also able to finalize two other wristbands, two men had come along and teleported beside them. It was Yvan and his son, Ysaac, volunteering for the mission.

The plan was simple: teleport to specific coordinates near the Tree of Life and wait for Harold and Nia to be transported to the Tree of Knowledge of Good and Evil (Patia reported that the prisoners were expected by twelve noon). Liam and the twins would create a distraction, then whoever could get to Harold and Nia would teleport them to the safe house. The team would reconvene at the safe house, stay for the night to make sure they weren't followed, then teleport to Casiah Desert for Pangea.

It was simple on paper, but everyone knew anything could

go wrong and throw them off the loop.

Exa smiled at her twin. Annisa had disguised them. She helped Alec dye his blond hair black and even styled it to give him bangs. Alec, for the first time, had put on black contact lenses. He looked so dark and different.

Exa refused to dye her hair. For some reason, it felt like her hair was the only natural thing she got at birth, given her anomalous eyes. Instead of her dark blue contact lenses, she exchanged them for hazel ones. Annisa gave her a makeover, dyed her eyebrows darker, and gave her a mole on her cheek.

Liam snorted when he saw her that morning. "What changed?"

Alec had to agree with him. But Exa was hesitant to change anything.

The textile tunnel of Pangea also gave them silver bodysuits, which they claimed had color-blending technology that could help them blend in the desert.

"Apollo, Artemis, are you sure you're up for this?" Liam raised an eyebrow at them. One of the rules was to use their code names.

"Yes, Harry," Alec and Exa said in unison.

Liam sighed and nodded. He had no choice but to trust the twins.

"There are the boulders Heph detected from before." Yvan pointed. "We'll hide there."

The four nodded and followed him.

Alec looked at his wristband. It was an hour to noon. Anytime now. From their position, they could view the big entrance of the Tree of Life. Its blue glass walls were situated obnoxiously in a tree trunk, blending with the tree branches.

Exa surveyed the desert. There were very few plants here but more boulders by their right. She could see the Tree of Knowledge of Good and Evil up ahead. It would take at least a five-minute walk in the desert. But Eden's sun wasn't scorching hot. In fact, she missed the pink hues of the sky above and the etching of the big pink moon.

In a different scenario, the twins would have admired the architecture of the Tree of Life and the Tree of Knowledge of Good and Evil, but not today. Today, the structures loomed hauntingly on the desert horizon, reminding them of their fugitive status.

Another thirty minutes passed. Yvan started to become impatient, sighing occasionally and checking his rifle and bullets.

Then, the holographic doorway of the Tree of Life glowed. Electronic static sounded briefly.

"Mom! Dad!" Exa whispered. She put a hand on her mouth.

Liam threw her a stern look.

They watched Harold and Nia, bound by red lasers around their wrists behind their backs, walk in front of at least twenty AI robots—the same robots that had invaded their home.

The last figure to exit the doorway was a man with a proud face and back. He had dark hair and a beard. He wore a dark cloak with a big golden apple pin.

"Unbelievable," Yvan whispered. "Agham Malaya will escort them? And twenty robots."

"Malaya?" Alec repeated, knitting his eyebrows.

Liam rolled his eyes at Yvan, annoyed.

Alec and Exa exchanged a worried glance. There were

only a few Malayas in Eden.

They watched Harold and Nia climb down the staircase. Harold nearly stumbled. There was clearly something wrong with his leg. His face looked bruised. There were burn marks on their wrists and arms.

Liam nudged Alec and Exa, who both nodded. The three scurried like rats to the boulders on their right.

One AI robot detected movement. It stopped and looked in their direction. The other robots continued marching, pointing their laser guns at Harold and Nia, who continued walking the desert path.

Agham noted the one robot. "What is it?" he barked and looked around, furrowing his brows. No one in their right mind would attempt anything. Twenty AI robots were programmed to shoot, kill, and burn.

"Detected movement on the outer periphery," the AI robot stated.

Agham's eyes narrowed, looking around more.

Yvan and Ysaac remained at the first boulder, holding their breath.

Liam, Alec, and Exa made a signal.

Harold and Nia had walked a few more feet, leaving behind the one AI robot and Agham Malaya for a little bit, when Liam suddenly stepped out.

He suddenly ran in front of Harold and Nia. "Get ready!" he shouted.

The AI robots suddenly raised their laser guns. Harold and Nia couldn't believe their eyes.

Agham saw Liam. "Get him!" he bellowed, running

toward his prisoners, followed by the first AI robot.

Liam grabbed Harold with his left hand and pressed his wristband, teleporting him and Harold out of the desert.

Yvan and Ysaac jumped from the boulder and began shooting at the robots.

BANG! BANG!

"Shoot for the nooks and spaces in their necks and joints!" Yvan instructed his son.

The AI robots turned their backs on Nia, who looked shocked, to target Yvan and Ysaac. Agham ran his quickest, trying to evade the bullets and lasers.

PING! PING!

He suddenly lounged for Nia. "YOU!" he shouted in rage. "YOU PLANNED THIS!"

"MOM!" Alec suddenly shouted, jumping out and running to Nia and Agham. He immediately pushed Agham, who lost his hold on Nia.

"Mom!" Exa shouted and ran to Nia. Her wristband was already set to teleport.

ZAP! At that moment, Liam teleported back. "Artemis! Apollo!" he shouted when he saw them in the middle of the ruckus. He ran forward.

Agham faced his transgressor with a new kind of rage. "YOU! WHO ARE YOU?"

Alec was surprised upon seeing Agham's face directly for the first time. His eyes. Black and almond-shaped. His cheekbones. The clench of his jaw. That stubborn and angry look. Violet Malaya, he immediately thought. Agham was Violet's dad. For almost eighteen years of friendship, they never saw Violet's dad. In fact, it was always Violet who went to their house, or they would always meet somewhere else.

Suddenly, Alec felt himself flying onto the sand as Agham's fist connected with his face.

Nia screamed. "NOOO!"

Agham suddenly realized. "Oh, is it you? You have changed!" He lifted his fist again.

"SON OF A BITCH!" an unfamiliar shout rang, and Alec was surprised to see Agham reeling backward onto the ground.

A rugged man with dark brown hair had punched Agham's jaw, saving him. He looked wild and enraged.

BANG! BANG!

Rifle sounds rang from the other side. Alec and Exa looked to see a blond woman shooting AI robots who had started surrounding Yvan and Ysaac. The twins stared. Like the rugged man, she also looked dirty and wild, with dust covering her golden hair and cheeks. But oh, she looked exactly like Exa. Only there were a few lines on her face that showed maturity, etched by time and hardship.

"I thought you'd never show up!" Yvan shouted.

"SHIT!" Liam joined the woman and started shooting AI robots.

The rugged man looked at Alec and Exa. "Fall back, the two of you!" he shouted.

"Behind me!" the blond woman called out.

Exa shook her head and turned to Nia. "No! We'll get Mom to safety first!" She turned to Nia and started to press her wristband.

"NOOOO!" Agham bellowed, and he lunged forward to

Nia, pushing her out of Exa's hold.

Nia shouted. "AHHHHH!"

ZAP!

Exa disappeared and teleported to the safe house but was unable to bring Nia with her.

"MOM!" Alec shouted as he scrambled to catch Nia. But Agham was quicker.

"NO!" the rugged man shouted. "You follow where she went and stay there!" He was pointing to where Exa had disappeared.

Alec felt himself burn with anger for a brief moment, at the thought of leaving their mother behind. No. Never.

BANG! BANG! PING! PING!

Twenty AI robots were not enough for five armed humans. They were deadly, yes, but too slow. Ysaac started laughing as he continued shooting at them like a maniac.

Agham realized this as he clutched Nia's body closer to his. He struggled to get up.

The blond woman suddenly pointed the rifle at him. "LET HER GO!"

Agham stood still, his left arm looping around Nia's right arm. Nia hissed and cried as the red laser continued to burn the skin of her wrist from all the movement. Agham smiled at them, then started to move forward, pulling Nia with him.

"You wouldn't dare," Agham sneered.

"We're warning you!" Liam shouted as he repointed his rifle from the AI robots to him.

But Agham's smile didn't falter as he continued to walk forward. "Are you ready to kill the woman who served as the twins' mother in your stead?"

Alec sucked in his breath as he looked at the blond woman. *No, don't*, he thought. Nia already seemed so weak from his viewpoint.

The blond woman sighed and hesitated, lowering her rifle a bit.

"Release Nia!" the rugged man shouted. "You are surrounded, Agham!"

"Am I?" Agham sneered as he stopped at the spot where Exa had disappeared. That was when Alec realized what he was up to.

"NO!" Alec shouted as he ran forward. Exa reappeared out of thin air at that instant, teleporting back when she failed to get Nia the first time. But Agham had already predicted this, catching Exa with his other hand and then—

ZAP!

"EXA!" Alec shouted as he stumbled into the empty space. He stared in horror as he realized he was too late. Agham had teleported with Nia and Exa.

For the first time, the twins were separated. His twin was captured by the enemy. Alec felt a deep regret in his stomach as the realization hit him.

Chapter 21

Something Is Fishy

The art of war is of vital importance to the State. It is a matter of life and death, a road either to safety or to ruin.

—Sun Tzu, *The Art of War*

Thirty minutes to noon, Azula was still happily showing Violet around the Tree of Knowledge of Good and Evil.

Violet had to clear her throat, knowing she must implement her plan now. "Mom, I hate to break it to you, but remember, I have to go to the Tree of Life after this."

The plan was simple. Meet Mr. and Mrs. July in the desert, on their way from the Tree of Life. They would have to walk because of the teleportation ban inside both Trees. Redirect the AI robots gently. Then, teleport Mr. and Mrs. July. She would deal with her father later. She had lots of excuses prepared inside her head.

Azula's smile remained. "Oh, no need, Violet," she said. "Your father will meet us here. He will be bringing the prisoners himself."

Violet stared. "Wait, what do you mean?"

"He is already on his way," Azula said.

"Well, uhm," Violet stammered. The first thing to go wrong in her plan. "Shall we meet him?"

"No need," Azula repeated with a tight smile.

Violet stared at her mother. Something was amiss. "Is he going to be all right?"

Azula's smile didn't widen. "He should be. Why? Do you think he'll be in danger?"

Violet thought fast for an excuse. "He's dealing with prisoners. So, I'm just a little bit worried—"

Azula waved her hand. "He'll be fine." She turned away, to another machine. "Now, this is another kind of genetic modification machine. It mimics the genes of certain animals, for example, birds, which we then input to another human temporarily, to grant a certain ability, like flight."

That momentarily caught Violet's attention. She looked at the machine. It looked decrepit, like a computer. Could it be the machine that Exa had seen in Isle Luna?

"Only flight?" Violet found herself asking.

"That's the first thing we experimented on. But we're trying other gene modifications, like a chameleon's," Azula smiled.

Violet made a mental note to return to this machine later. "Mom," she began. "It's almost twelve. Should we check on Dad?"

Azula sighed and looked at her wristband. Then her eyebrows knitted, looking at the red dot. "Hmm, they seem to have stopped halfway. It must be something minor."

Violet gasped. "Or something major, Mom!"

She turned around and ran.

"VIOLET!" her mom shouted after her. "Where are you going?"

But Violet had run out of excuses. If her father and the

prisoners had stopped halfway, something must have happened. It might be easier for her to get to Mr. and Mrs. July. Or more complicated, since this wasn't part of her plan.

Violet ran out of the sphere, heading toward the holographic exit of the Tree of Knowledge of Good and Evil.

"Violet! Wait up!" Azula ran after her. Other scientists turned to watch the ruckus.

Violet cursed under her breath. Oh no, not Mom. She ignored her mother as she ran out of the tree. She ran down the stairs and heard the gunshots.

BANG! PING! BANG!

Oh no, Violet thought in horror. Just ahead she saw several people. AI robots lay broken on the sand. Two men were shooting them. She saw her father holding Mrs. July…and Exa? Mr. July was nowhere to be found. Alec was running while two other people were pointing rifles at her dad.

"Dad!" she shouted. "Alec! Exa!"

Then, her dad vanished with Exa and Mrs. July. Violet gasped and stared. He had teleported.

But to where?

At that instant, her eyes met a familiar pair. Cold, brooding, but familiar. The same eyes that were gazing at her just a few days ago, when they were watching seawater violets bloom in the sunset. Alec's. He had dyed his hair dark. A disguise? He could never fool her. He looked anguished. He was panting. He looked like a madman. This was the first time she had seen him like this. The twins always looked calm, collected, even aloof. Violet felt an ache for him she could not ignore. She called out his name.

"Alec, Alec," she repeated, her voice coming out in short

raspy breaths after all that running. She took a few steps forward. To hold him. To be there for him.

"STOP!" A hand suddenly grabbed her, forcing her to turn around. It was her mother.

"What are you doing?" Azula shrieked in front of her face. "Let's go!" She had seen the commotion and her husband teleport.

"Mom, they need help!" Violet reasoned.

Azula's eyes narrowed. "Do you realize what you're saying? Any idea at all? What does this mean to us? Your dad? Our family?"

Violet stopped struggling.

BANG! BANG!

Mother and daughter saw the last AI robot tumble in electric chaos onto the sand. Alec and the other people were watching them now. His face contorted in sorrow and confusion.

Violet couldn't stop her tears as they fell. Azula's face was stone-cold as she stared ahead at the intruders.

"I'm sorry, I'm sorry," Violet called out to her friend.

Azula pulled her daughter and dragged her to the stairs to the Tree of Knowledge of Good and Evil.

"I'm sorry, I'm sorry…" Violet couldn't stop her tears. Alec watched her being dragged away, pained. A look of confirmation now in his eyes.

A blond woman with a rifle put her hand on Alec's shoulder. They all stared at mother and daugher. Another man pointed his rifle at Azula. But a big man shook his head and motioned for him to lower his weapon.

Azula and Violet entered the Tree of Knowledge of Good and Evil. Once inside, Violet couldn't help but to sob even more.

"Let's go to my office," Azula said coldly.

The other scientists started to look curiously after them, but Azula waived their concerns, giving small smiles as she and Violet passed them. When they reached the top floor, Azula's face hardened.

Violet tried to wipe her tears and wordlessly followed her mom to one of the spherical rooms.

Finally, when they were alone, Azula spun around to face her daughter. "What were you thinking, Violet? Is this why you wanted to meet your father? For what?"

Violet looked down, not saying anything. She sniffed.

"You have to remember who you are!" Azula exclaimed. "You are our daughter! What would your father say? Were you planning to free the twins' parents?"

Violet bit her lip.

"Did you even think about your plan?" Azula shrieked. "We raised you better than this, Violet! At least make it foolproof!"

Violet proceeded to cry silently.

Azula continued. "I know they are your friends. I even talked to your dad about this. But there's nothing we can do now. You will be taking a position in the Tree of Life Council when you're older. You are our legacy. You cannot foolishly run around like this. Do you understand?"

Violet nodded.

Azula sighed. "I know your father. He will be all right. But please, don't make me worry about you too. This is a bigger issue than you'll ever know. I think, more than your dad knows."

Violet looked at her mother. "Do you know something more about this?"

Violet thought she saw hesitation in her mother's face, but it was gone, and she wondered if she had really seen it the first time.

"Of course not," her mother answered. "I'm just a scientist. What should I know?"

"The head scientist," Violet corrected her.

"You have the best of both Trees. Your father and I can give you so much," Azula said, walking toward her. She put her arms around her daughter's shoulders. "Play your cards right, and don't throw these opportunities away. Your future is bright, Violet. I have made it so. All right?"

Violet forced herself to meet her mother's eyes and nodded.

Azula hugged her daughter. "I'm always thinking about you."

"Mom?" Violet whispered. "What about Dad? Exa?"

Azula had an idea. "They will be here soon. But perhaps not today, after all that commotion."

Chapter 22

Alexander and Miriam Leons

But when she could hide him no longer, she got a papyrus basket for him and coated it with tar and pitch. Then she placed the child in it and put it among the reeds along the bank of the Nile. His sister stood at a distance to see what would happen to him.

—Exodus 2:3–4

They all reconvened at the safe house, an abandoned stone house on Isle Akalasa, a snowy floating island with only a hundred inhabitants.

Alec found himself bent over a kitchen sink. He washed his face, almost as if trying to wake up from this nightmare.

"At least we have you back," Liam said, hugging his old lost friend, Harold, as they entered through the kitchen door. Harold still had his wrists bound with red laser behind him. His skin was reddened and zinged, and there were even a few droplets of blood.

Alec could only imagine their mother's wrists. They were probably in a worse state.

No, Alec angrily thought. How could he even face his father right now? They weren't able to save Mom. They even lost Exa.

"What do you mean?" Harold's panicked voice filled the air. He had spotted his sister. "Miriam! Thank the heavens you're here! Where is Nia?"

The room fell silent. Alec didn't even want to turn around.

"AGHAM GOT HER?" Harold's voice rose when Liam told him. "AND EXA?"

Alec cursed as he faced his foster father. He crossed the room.

Harold saw his state. "Alec, we will—"

"I'm sorry." Alec's voice broke. "I didn't get there on time. Agham saw what would happen first."

"No!" the rugged man and blond woman suddenly objected.

"It wasn't your fault, Alec." The blond woman put her hand on Alec's shoulder.

Just like before, Alec shrugged it off. And like a while ago, in the desert, the blond woman looked dejected.

"Alec," Harold began. "You did nothing wrong, son."

Alec looked pained, and then his eyes began to tear up. "I couldn't protect her, Dad. I failed to! I didn't even save Mom!"

"You did your best, Alec," Harold insisted. "Agham Malaya is a cunning man—"

"You know him!" Alec said. "He's Violet's dad!"

"Yes," Harold said. "That's why we have never let you and Exa go to Violet's place. And why we just invite Violet over. So that Agham will have less chances of seeing you."

"But we have always been so careful and watchful of you and your sister," the blond woman, Miriam, said softly.

Alec didn't say anything.

"Son," Harold said. "I know we should've told you and Exa earlier. And we really planned on telling you after your eighteenth birthday." He paused, glancing at Alexander and Miriam. "These are your parents, Alexander and Miriam. Miriam is my sister. I don't how much Ajax told you—"

"All of it," Liam said.

Harold smiled woefully. "Well, he beat me to it, then." He turned to Alec. "I hope I have been a good dad to you and your twin, albeit for a short time—"

"You will always be my father," Alec said, then embraced him tightly.

Harold smiled. "I would hug you back, but these lasers are burning my wrists."

Alexander and Miriam looked on wistfully. Alec hadn't acknowledged them yet.

"Heph might have something in Pangea for that," Liam said. "I honestly don't know how to disable those lasers."

"We'll be home in the morning," Yvan said. "We must follow protocol, Harold, just in case we were tracked. Especially when Agham escaped."

"I understand," Harold answered. "But I don't want to leave without Nia and Exa."

"We have to go back!" Alec announced loudly. "Chances are he took them somewhere else, like another safe house, before returning to the Tree of Knowledge of Good and Evil. He will have to bring them there somehow."

Harold nodded. "Agham wanted to test the Memoir Edit, as he calls it."

"I also doubt he would just let Exa go, or kill her," Liam said. "He will find ways to study her genetic makeup."

Miriam cursed.

"We will go back for them," the rugged man, Alexander, said.

"Agreed—" Harold began.

"But it would be easier to save them with free hands, right?" Liam said to Harold. "You will go to Pangea first. Recover—"

"No!" Harold interjected. "I have to get Nia before they wipe out her memories!"

"We will not let that happen, brother!" Alexander said. "We will stand guard for Nia and Exa."

"You can go ahead with my brother by tomorrow's first light," Miriam told Yvan and Ysaac.

"I will do this mission with you," Liam told Miriam. "Exa was also my responsibility."

Alexander's tone became accusatory. "I am surprised Ajax let my children join—"

"We wanted to," Alec interjected. "It's our parents!"

Miriam and Alexander exchanged a glance.

Harold sighed and looked at Alec pleadingly.

Alec bowed his head. Then he said softly, "I will join you tomorrow too."

"Alec!" Harold and Miriam groaned.

"Kid," Alexander began. "It would mean a lot to us if you were safe! You have no idea how much we sacrificed to give you your life now!"

"My life now?" Alec said, his voice rising.

"Alec," Harold sternly warned.

"Well, I knew we were hiding Exa's heterochromia, but imagine suddenly knowing there was another world underground, that I have another set of parents, and suddenly my best friend's dad is the enemy!" Alec thundered. "And I just lost my twin. She's been the only thing constant ever since!"

Alexander opened his mouth, but Miriam stopped him. She took a step closer to Alec. "Alec," she began. "I know it's a lot to take in—"

"It is," Alec seethed.

Harold sighed. "Alec, it hasn't been easy for us either."

Alec suddenly looked ashamed.

Miriam sighed. "Okay, we will let you go with us tomorrow—"

"Miriam!" Alexander hissed.

Liam groaned.

"—but you are not allowed to leave my side, okay?" Miriam said.

Alec looked a bit relieved but still angry. He nodded.

"I guess this isn't the family reunion you were looking forward to," Yvan snorted.

Alec looked away. Alexander, Miriam, and Harold exchanged somber looks.

Harold cleared his throat. "Don't worry, we will get Nia and Exa, Alec, I promise."

Ysaac suddenly asked, "How come you knew where we were? We weren't expecting you." He was looking at Alexander and Miriam.

Alec looked at them. He had also been wondering about that. Why did they show up suddenly?

Alexander and Miriam exchanged a look. "We…stole something." Miriam cleared her throat.

"That wouldn't be the first time." Liam shrugged. "We have been scavenging for scraps in Eden for a long time."

"From Edmund Siegart," Alexander finished.

"WHAT!" Yvan exclaimed, his eyes widening in surprise. "For what! Oh, wait until the other elders hear about this!"

Even Harold looked at them in disbelief. "Why? What did you take?"

"Have you realized how risky that was?" Yvan asked, his voice rising. "EDMUND SIEGART! The richest council member!"

"We were planning to save you," Miriam said, pointing at her brother. "And Nia."

"When we reached your house, we couldn't find the twins," Alexander added. "We figured they were able to escape. We thought it would be better for their safety if we saved you and Nia first."

"Don't worry," Harold said. "Nia and I knew not to talk."

"I dropped my tracker back at the house, because I thought the others might come looking. And, well, we couldn't enter the Tree of Life without the golden apple pin." Miriam sighed.

"YOU STOLE A GOLDEN APPLE PIN?" Yvan's voice rose again.

Even Alec knew that move was too risky.

"We've already been wanted for eighteen years!" Alexander rolled his eyes. "It's not like we killed someone. We just stole it."

"Were you caught?" Liam said. "That makes a difference, right?"

"Nearly," Miriam admitted and sighed.

"His island," Alexander emphasized, "has AI robot guards."

"As expected from a rich council member," Yvan scoffed. "Of all council members, you had to choose Edmund Siegart! You could have chosen Rosarius Sancti! The guy's so fat I wonder why he hasn't had an AI liposuction yet. Rosarius would never notice you stealing his pin when he's snoring at night!"

"We were at the right place at the right time," Miriam said as a matter of fact. "He was the first council member we saw who exited the Tree. We tailed him."

"Then tackled him," Alexander sighed.

"TACKLED HIM?" Yvan exclaimed. "So he saw your faces!"

"He panicked, teleported, and brought us with him in his private island. We teleported into an observatory tower," Miriam said.

"A what?" Liam, Yvan, and Ysaac said in unison.

"He's planning something," Alexander said. "He has an AI telescope at the top of the tower with several documents labeled Galileo X? I don't really understand it. I don't really care. Maybe he was searching for a planet or something up in the skies."

"We were able to snatch the pin," Miriam admitted.

"Several AI robots chased us, and I guess they were able to ID us. But we were able to escape. Emergency parachutes are really important." She sighed.

"First time spotted in eighteen years," Harold said. "Hmm, I guess that's not too bad."

"We'll worry about Edmund Siegart later," Yvan decided. "We'll report it to the other elders. So now, you have a golden apple pin we can use to enter any of the Trees."

Alexander and Miriam smiled. "Yes."

"Then what are we waiting for?" Alec said. "We could save them now!"

"Whoa, whoa, whoa!" Harold, Yvan, and Liam all tried to subdue Alec.

"We still don't have a good plan!" Yvan said.

"You can't just storm in there," Liam said, rolling his eyes.

"Always have a plan A and a plan B!" Harold cautioned.

Alec sighed angrily, but he let them talk.

"First things first," Yvan began. He looked at Harold. "Ysaac and I will take Harold to Pangea and get Heph to remove those laser things. We will lend you one of our rifles." He looked at Alexander. "Then Alexander, Miriam, Liam, and Alec, please remember your code names, all right? You will head back to the Tree of Knowledge of Good and Evil. Our target is the Memoir Edit because that is where Agham plans to take them."

He paused. "But I want someone to stay and watch outside the Trees just in case."

"Guarding outside?" Liam repeated.

"Yes." Yvan nodded. "Chances are Edmund Siegart has

already reported his stolen pin. Agham will be notified, of course. He might change his plans. Would he place the Memoir Edit machine inside the Tree of Life Council? Or his home? I don't know. Or increase the security of the Trees."

"The guard outside will be in constant communication with the ones inside," Yvan continued. "Those inside will have one goal: find Nia and Exa and bring them out. It will be difficult, seeing we don't know what is in there—"

"We could ask Patia," Alec suddenly said.

The others started nodding in agreement.

Yvan sighed. "All right, then you all have to wait for us to get to Heph and for Heph to ask Patia before you go in."

"Don't we have any direct communication with her?" Alec asked.

Liam shook his head. "Only Heph and Ajax talk with her directly."

"Not even with us, the other elders." Yvan smiled. "It's for Patia's safety. She had to cut connections with almost the whole of Pangea when she infiltrated Eden."

"Yes, that's for safety." Miriam smiled, looking at Alec. Alec looked away.

"Wait for Patia's instructions, then," Harold said. "That way, you can wait for me to come back. I will join the mission too."

Yvan rolled his eyes. "There's too many already!"

"You can stay in Pangea, Yvan," Harold said. "We are too many, yes, but this is family."

Yvan sighed. "Fine. I guess that's the plan. Hopefully, Patia will help us with the mission tomorrow."

Chapter 23

Council Problems

Lola said You must learn to swallow
your heartaches, *I was ten, knew nothing
of love, save for presents: Coloring*

books, paintbrushes.
I had the precision: fingers drafting sprinkles and spires on eggshell paper.

—Alyza Taguilaso, *Pearl*

"Mom!" Exa shouted as she felt the air around her tighten and throw her out into another place.

Nia gasped beside her and fell to the floor. Her wrists were already burnt, oozing with pus and blood. "Exa!" she exclaimed. "Are you all right?"

"I should be asking you that," Exa said as she struggled to help Nia up.

"You!" a sharp voice said, resonating. They looked up to see the man who had brought them out of the desert: Agham Malaya.

"They have nothing to do with this," Nia said, starting to pleading. "She was just a baby!"

"Not today she wasn't," Agham said. "Or her twin."

Exa saw Violet in Agham's face. She sighed, wondering

why they had never made the connection in the first place. Why Violet never told them her dad was a council member.

"What are you planning to do?" Exa demanded. "Stay away from us! Haven't you done enough?"

Agham calmly looked at her. "And haven't you done enough? Sabotaging Council plans, helping prisoners escape, destroying Eden property, and more."

"You attacked us first!" Exa seethed. "You burned our home! Hunted us down!"

"As you should be," Agham merely said. "I've told your aunt. Wait, you already know? Or not?" He looked at Nia.

Nia glanced somberly at Exa. "I'm sorry—"

"It's okay." Exa turned to her and reassured her. "We already know."

"Good," Agham said. "Less recap for everyone. But again, Exa, you shouldn't have lived in the first place. Mutations are not allowed in Eden."

Exa sneered at him. "That's what you tell yourself at night to sleep? Stop covering Project Minus with that law!"

Agham's eyes narrowed. "How did you know about that? You're good." He sneered at her, looking at her from head to foot. "But the burden of making Eden a paradise doesn't lie on your shoulders. You don't have a say."

"Where are we?" Nia suddenly demanded. She looked around. They were surrounded by white walls.

"This is my home," Agham angrily admitted. "See how hospitable I am? This was the first place I thought of. You, however, cannot teleport in or out of my house. I have reprogrammed the space here."

Violet, Exa thought. Was she here? Could she ask for help?

Suddenly, Agham's wristband blinked red. He sighed. "You will stay here until tomorrow. There will be AI robots patrolling. You will be foolish to try to escape."

With that, he walked away and exited the room.

Nia scoffed. "I don't think Agham is used to having much action for a day. The Tree of Life Council, ugh!"

Exa suddenly hugged her mother. "Are you okay, Mom? What did they do to you? I saw Dad. He also looked terrible."

Nia looked at Exa and smiled. "Exa, shhhhh. He will be watching us and listening to us."

Exa looked around. To her horror, she found an AI camera in the corner.

"Cardinal rule: never talk." Nia smiled.

Exa looked sorrowfully at her mother and hugged her.

"At least I can sit on the floor here," Nia sighed as she proceeded to.

"I'll try to find a way out," Exa promised.

Nia smiled sadly and shook her head. But she remained silent.

Agham sighed as he exited the room. "Do not let them out of this room," he ordered two AI robots as he passed by them.

He took the stairs from the basement. Their family library greeted him. What a day. And now Edmund was calling an emergency meeting this instant?

He answered with his wristband. "What?" he demanded.

Holographic figures of the other council members appeared. Edmund looked distraught.

"Agham! Agham!" he exclaimed.

"Why do you look like that?" Agham asked. The richest council member looked sweaty and flushed as if he had just run around.

The other members started grumbling. "What is going on? Couldn't we schedule this meeting? What is the emergency?"

"I hope you have a good reason for this," Agham sneered. "I was in the middle of an important business."

"My golden pin has been taken from me!" Edmund shrieked.

"WHAT!" Several shouts from the other members echoed.

"How? How did it happen?"

"Why?"

"WHO DID IT, EDMUND?" Agham demanded.

"I was able to identify the thieves!" Edmund said in a panic. Thievery was extremely rare in Eden. Everyone was rich and in abundance. There was no reason to steal.

"Who?" Agham repeated.

"AI identified them as Case Number 2," Edmund said. "Alexander and Miriam Leons."

There were several gasps.

Other members started complaining. Rosarius said, "What is going on, Agham?"

Levida said, "I told you before. Because those twins are involved, one thing leads to another. They must be after the prisoners."

Agham had to agree with Levida. "When exactly did they steal your pin?" he asked Edmund.

"Just a few hours ago. I am still trying to calm myself! They attacked me just outside the Tree of Life! I teleported and somehow, brought them with me. But I had AI robots catch them," Edmund said. "But they escaped! This is a nightmare!"

"Increase your security or find another island, Edmund," Agham instructed. "Your security there has already been breached. I suggested that the other members also be vigilant. We will tighten the security surrounding the perimeters of the Tree of Life and the Tree of Knowledge of Good and Evil."

"What about the prisoners?" Levida asked.

"Alexander and Miriam appeared just as we were transporting the prisoners to the Tree of Knowledge of Good and Evil," Agham informed them.

There were several gasps.

"How did they know, Agham?" Rosarius demanded.

"I was thinking the same thing," Agham said. There had been other unidentified people there—three men with rifles, accompanying the twins. Could they possibly know his plan about the Memoir Edit? How? Or were they with Alexander and Miriam, and just waiting for them to arrive with the stolen pin to infiltrate the Tree of Life?

"I will get to the bottom of things," Agham said. "But exercise caution! The possibility of a spy, a snitch, is not ruled out!"

"How about your daughter, Agham?" Jet, the youngest council member, said.

There were several murmurs. Agham clenched his jaw.

Levida spoke the loudest. "I agree, Agham! Your daughter has been identified as a close companion to the twins since they were young. Imagine the twins being in your midst, and you missed it—"

Agham saw red.

"—and I'm sure there is emotional baggage attached to your daughter—"

"MY DAUGHTER HAS NOTHING TO DO WITH THIS!" Agham thundered. "IN FACT, SHE HAS STARTED HER TRAINING TO BE PART OF THE COUNCIL! SHE HASN'T SEEN OR TALKED TO THE FUGITIVES SINCE!"

Everyone fell quiet.

Levida cleared her throat. "We're just saying—"

"HOW ABOUT YOUR SON, LEVIDA?" Agham shouted. "WASN'T HE THE ONE WITH THE TWINS THE NIGHT BEFORE? AND HE EVEN INVITED THE GIRL TO YOUR ISLAND!"

There were several murmurs of agreement. Levida looked taken aback. "Now, I'm willing to put my son into the Memoir Edit—"

"MAYBE YOU SHOULD! RIGHT AFTER I'M DONE WITH THE PRISONERS!" Agham barked.

"Where are the prisoners?" Jet suddenly inquired, breaking the tension over offspring.

"Harold July escaped," Agham bitterly informed them. "But I was able to personally capture Nia July and…Exa July."

There were several murmurs again.

"Once again," Rosarius said, clapping. "You are the best among us, Agham! Exa July!"

"Her genes must be examined!" Edmund shrieked. "And killed! After what her parents did to me!"

"Her genetic composition will be studied by all means necessary," Agham reassured them. "She will be the very example of why mutations are not allowed in Eden."

Chapter 24

Start Them Young

There were far too many
child prodigies. My paintings mere blister dreams
in an ocean of Promil Kids. They decided

I would be a doctor: never to starve, well-
fed by ambitions, tumors, and boils. Scalpel and suture
in place of pencil. I take it, gulp it down

like a pill. Gorge on. Pray
someday I might cough up a pearl
or choke.

—Alyza Taguilaso, *Pearl*

ZAP!

Violet pulled down her hood to cover her face more as she teleported back to Isle of Templos. There was the usual noise of rowdy teenagers partying until the next day. She quickly moved away from the crowd and headed toward the temple ruins, deeper until the party noise faded.

"Theos," she spoke into the darkness.

Theos appeared behind her. "What took you so long? I messaged you this morning."

"Don't tell me you have been waiting for a day here,"

Violet replied. "I told you I'd come at dusk."

Theos sighed. "Of course not. I…" He looked away. "I asked my mother if I could also start my training."

"And?" Violet looked up.

Theos took a deep breath. "She doesn't want me to. Not yet, she says."

Violet shrugged. "That's unusual. Council members usually want early training by the time their children reach eighteen."

"You're seventeen," Theos noted. "But you act like thirty."

Violet raised an eyebrow.

"Well, the twins too," Theo said, looking away.

"I guess we just think alike in so many ways," Violet sighed. "So, what will you tell me?"

"The Memoir Edit," Theos began. "I'm sure you've heard of it."

Violet nodded. "My mom showed me the tech a while ago. It's a big circle, and it has a bed in it. You have to put on a steel headgear. The rays will focus on your head, and memory modification will commence. It will only take a few minutes. Depends on how many memories will be edited."

Theos nodded. "My mom wanted us in there." He looked sheepish.

"WHAT?" Violet exclaimed. "Because we know the twins?"

Theos nodded.

"And you're going along with it?" Violet asked him incredulously.

"Of course not!" Theos smirked.

"Really?" Violet teased him. "Aren't you little mommy's boy?"

Theos looked offended. "Hey!"

Violet suddenly laughed. For the first time in days, it felt good to laugh.

Theos looked at her as if she had grown another head. "That's a first," he quietly noted. "You should smile more."

Violet rolled her eyes and growled at him.

"Now you've ruined it." Theos smirked. "You look like Agham now. Who's daddy's little girl?"

Violet rolled her eyes again. "For the record, I was the one who suggested early training. I was hoping to find facts about the cases and the twins." She sighed. "The Council has been keeping more and more secrets."

Theos nodded. "Yes, they are. I believe you have only uncovered the tip of the iceberg. You told me about the twins' real parents, right?"

Violet nodded.

"I have another news," Theos said. "Their biological parents were identified stealing Edmund Siegart's golden pin."

"What?" Violet gasped. "What for?"

"What do you think?"

Violet nodded. "Oh, to save Mr. and Mrs. July. Luck's not with them. Or us." Then she told Theos everything she had witnessed that afternoon.

"Now your father's got Exa," Theos groaned.

"I have an idea of where he took them," Violet said calmly.

"Don't worry, they should be safe. Or else I wouldn't have met you here. Instead, I would've made my way to them."

"How certain are you?"

"There are only a few places my father feels safest." Violet shrugged.

"The Memoir Edit," Theos reminded her.

"I plan to sneak out Exa and her mother tonight," Violet said. "Before they get them into the Memoir Edit. It should be easier now."

"Again, how sure are you?" Theos asked her. His eyebrows knitted together.

"I'm pretty sure my father took them to my house." Violet smiled at him.

"Oh," Theos said. "Can I help?"

"Nope!" Violet said. "I have something for you. You will find a way we can avoid the Memoir Edit."

"It will be hard to," Theos said sheepishly, looking down.

"What is it now?" Violet rolled her eyes.

"I overheard the emergency Council meeting earlier when Edmund reported his lost gold pin. My mom seems to be sowing doubts about your dad. And there might be a spy inside the Tree."

Violet's eyes widened. "This is bad. If everything worsens, my dad will have me and you in that machine."

Theos nodded.

"We have to take precautionary measures." Violet decided, then frowned. "Do you have any ideas, Theos?"

"A reminder," Theos said. "A note—"

"Will be too risky," Violet finished. "A simple note can be traced to us. No one should know but us."

Suddenly, Theos flashed a smile. "Are you open to body art?"

Violet stared at him, incredulously. As the realization dawned on her, she let out a small groan.

ZAP!

Violet entered her house. It was big, yes, but not as big as the other council members'. Isle Luna was practically Levida and Theos's playground. Edmund Siegart reportedly owned at least three private floating islands. Rosarius Santi had a small private island with towering houses. Jet, well... Violet frowned. The youngest council member was said to be rich, but he also liked to keep things quiet, like her father.

This house was nothing compared to what Agham and Azula could do or achieve. Heads of both Trees. Everyone in the Council knew who really owned Eden.

Casting doubts about her father was a big, bold move, Violet thought. She had to admire Levida Luna's tenacity. And she hated it as well.

Levida Luna had an interior motive, Violet realized. But what? She pushed the thought behind for now. First things first. She had to check the basement. She had a hunch she would find Exa and Mrs. July there.

But as she made her way to the library, she heard the clanging of silverware in the direction of the dining room.

Violet paused. It was early evening. Was it really true? Her parents had come home early? Violet thought of her mother earlier that day. Would she tell on her to her father? She decided to enter the dining hall. She opened the doors and found her parents eating dinner. Violet blinked at the sight.

They looked up at her as she entered.

"Violet," her father greeted her as he gestured to an empty plate beside her mother's. "Come join us."

"Where were you?" her mother asked. "You left earlier than I did. I was hoping to find you here."

Violet shrugged. "Just wanted to watch the sunset."

"The sunset?" her father repeated as he munched on his food. It was as if he hadn't taken on several "fugitives" that morning.

"The sunset and the saltwater violets." Violet smiled. "It's been a while since I watched them bloom."

Her mother and father exchanged glances.

"Of course," Azula said. "It is healthy to take a breather once in a while."

Her father said, "I am proud of you, Violet. Your mother told me you learned much at the Tree of Knowledge of Good and Evil."

Violet glanced at her mother, who winked at her. She smiled at her father. "Yeah," she said, trying to fake enthusiasm. "It was fun! I didn't know we had all that tech there!" She helped herself to some dinner.

"I'm glad," her father said. "Tomorrow, you will spend the day with me."

Violet stared. "With you? All day?"

Her father nodded. "I told the Council that you will join me, except for Council meetings."

Violet nodded. "All right, I'd be honored to."

Then her father sighed and announced, "Tomorrow, you will join me for the rehabilitation of our prisoners."

Violet felt a cold sinking feeling in her stomach. "Rehabilitation?" she repeated.

Her father smiled. "I will show you tomorrow. We have developed a technology that will allow more humane punishment of prisoners and let them rejoin society again."

"The Memoir Edit," Violet whispered.

"Good!" Azula beamed at her. "You remember!"

Agham looked at Violet straight in the eye. "You are familiar with Nia July, right?"

Chapter 25

Family Affairs

If you cannot get rid of the family skeleton, you may as well make it dance.

—George Bernard Shaw, *Immaturity*

Dinner went on, as normally as it could. All issues swept under the rug.

Later that evening, when Violet was certain that her parents were sleeping, she silently made her way to their basement. Her father had his own entrance to the basement via the library.

The usual AI caretakers did not take notice of her when she passed by, but she suspected that there would be guards downstairs.

When she made her way to the basement, white lights automatically turned on, signaling her presence. Violet continued on. AI robots looked up at her.

"Agham Malaya asks you to let the prisoners go," Violet announced.

"Code requested," one of the AI robots said. "Voice doesn't match."

Violet smiled. "1-3—"

"I knew you'd be here," a familiar voice interrupted Violet.

Violet gasped, turning around so quickly that she almost tripped. It was her mother descending the stairs.

Azula smiled. "Violet, we have talked about this."

"Mom," Violet whispered. "Please."

"Did you really think your father would let you release Exa and her mother?" Azula said. "I know they mean a lot to you. But think of your future, Violet."

"They don't deserve to be under the Memoir Edit—"

"I think it's wonderful," Azula exclaimed. "Imagine, Nia can join Eden again without the burden of the memories of her taking care of fugitives. She can live without guilt."

"No, no," Violet whimpered. "Exa—"

"She is a different case. Her heterochromia needs to be studied," Azula announced. "A study she managed to forego when she was born."

"Project Minus—"

"And how did you know about that?" Azula narrowed her eyes. "Has your father told you about that? I haven't. Oh, well, you will know soon enough. You will learn soon enough."

"Learn what?"

Azula smiled. "The dots will connect for you in no time, my darling. You'll understand that there is no black and white. Most of the time, you have to choose the lesser evil. You will understand soon."

But Violet became more confused. "Please, Mom, I have to save Exa and Mrs. July!"

Azula suddenly looked stern. "You don't know what you're asking, Violet. And in our own home. With your father upstairs. How many times have I told you to play your cards right?"

Violet stammered. "But—"

"No buts!" Azula said as if her daughter had regressed to being a toddler. "Go up to your room."

Violet took a deep sigh. "Exa didn't even know I tried," she whispered.

"And let it remain that way," Azula said. "It's best you don't associate yourself with them."

Violet took small steps back to the stairs.

"GO!" Azula barked. "And don't make me catch you here again!"

Crying, Violet scurried off, defeated. Now, she had to think of other means. Her mother was not her ally; she must keep that in mind.

Azula was left in the basement. She turned to the AI robots. "Now," she whispered. "Where were we?"

In a few seconds, her husband's shadow filled the staircase. "You were right. They were right," he sighed. "Violet's memories need to be modified as well."

Azula nodded. "It's for her own good. We can't have the Council turn on you. It's also for her future."

Agham nodded. The AI robots let him and Azula enter the interrogation room.

"Mrs. Malaya!" Exa gasped. She had only met Azula a few times at school events. She remembered Azula offering her and Alec cupcakes at a sporting event when they were young, supplying extra water during one academic competition, or just taking Violet to a park to play with the twins.

"Azula." Nia's eyes narrowed.

Azula smiled sweetly. "I'm sorry that it has come to this. And I've known you and the kids for a very long time."

Exa stood in front of Nia. "No!" she shouted. "I will not let you or anyone take her away from me!"

"Sweetie," Azula said. "You just have no choice."

Then Agham and Azula put on gas masks. Agham raised a hand. "Release."

AI robots suddenly released a gray noxious gas from their guns. Nia and Exa began coughing.

"Don't breathe!" Nia wheezed. But with the lack of food and proper sleep during her imprisonment, in addition to the wounds her wrists have acquired from the lasers, and her body nearly giving up, she was done for.

Exa coughed and struggled as she watched Nia lose consciousness. "MOM, NO—ahem! Ahem! Mom! Mom!"

Within seconds, mother and daughter lay on the floor, unconscious. Azula and Agham hadn't taken off their masks yet.

"Carry them," Agham ordered the robots.

Within seconds, they had all teleported before the Tree of Knowledge of Good and Evil.

"Our work begins," Azula announced.

Violet gasped in horror when she returned just an hour later to the basement. The room was empty and reeked of noxious gas. She covered her nose as she ran back upstairs.

She got dressed, wondering who to call for help. She thought of Alec. He had been wearing a wristband in the desert a while ago. Could she send a message?

It was worth a shot.

Agham's wristband alarmed briefly. He checked it. There was a notification.

Exa was laid out and strapped in a silver floating capsule. She was directed to the examining room. Nia was also placed in a similar capsule, but to be taken to a different spherical room.

Agham frowned. "Someone tried to message Alec July."

Azula met his eyes. "Who?"

"Violet," Agham sighed.

Azula grunted. Their daughter was proving to be more complicated than they expected. "Perhaps because I told you not to be too harsh on the twins and their parents," she said.

Agham turned to his wife. He raised his voice. "And then how do we get what we want? We are fortunate that Miriam and Alexander showed up."

Azula sighed and nodded. "Let's work with Exa first. Give me a couple of minutes to prepare."

"Find something. We must have something other than Project Minus to show," Agham said.

"I know what I'm doing, husband," Azula said. "How about you work on the other 'fugitives' you saw earlier?"

Alec sat on the windowsill, staring at the giant pink moon above. It had only been days since he, Exa, and Violet were gazing upon it aboard the *Voguer W*.

Somewhere, he could hear Liam snoring. Poor Harold was sleeping on the table, face forward, careful about the lasers that still bound his wrists. His biological father, Alexander, was chatting with Ysaac and Yvan.

He heard her footsteps before her voice.

"Alex," she said softly.

Alec sighed, refusing to look.

Miriam slowly approached his side. "I know you're still confused—"

"You left us," Alec interrupted coldly. "Had them lie to us all these years."

"It was for your own good—"

"Exa and I have been very careful almost all our lives. We already knew what was at stake," Alec said. "But you never even gave us a chance to get to know you."

"Nia and Harold were about to tell you," Miriam softly said. "But you're right. We should have told you sooner."

Alec remained quiet. Miriam took this as a cue to continue.

"We did that because we wanted you and Exa to have a chance to live in Eden. Pangea is wonderful, but everything has to be done in secrecy. It's not a place you would want to raise a family in." Miriam sighed. "I'm sorry, I'm sorry. I regret that we didn't catch you or Exa during your first steps. We weren't the ones who taught you about wristbands, went with you to school, or to meet Violet in the park or someplace."

Alec felt a tear on his right cheek. But he didn't move. Or even say a word.

"I'm sorry," Miriam repeated. "But we saw it all. Believe

me. When Harold took you and Exa to the mimosa fields by the island to run. You stumbled so badly that you wouldn't stop crying. When they tuck you in bed. Did you know it was my idea to have my brother make that passageway? Because I wanted to watch you sleep whenever I could. You and Exa. Your father and I were there. Those little fights you had with Exa over petty things? The first to finish a meal? Who left dirty laundry at the sink?" Miriam laughed a little. "I saw all that too. I'm sorry we didn't tell you sooner."

Alec sighed deeply. He knew his mother was telling him only the truth. There was no other way for him and Exa to survive. But it didn't diminish the ache in his heart; it was as if they'd been cheated for the majority of their lives.

"Alec—" his mother began, but she was interrupted by Yvan's booming voice.

"Heph has notified me!" he announced. "He'll be sending Patia's instructions."

"That's good news!" Alexander said. "Then we just follow them tomorrow. It will be easier."

Alec turned to his mother. "I should rest. You should too."

"Alec…"

But Alec left her abruptly and went upstairs to one of the bedrooms.

Alexander stood up from his chair and placed an arm around Miriam. "He'll come around. He has to."

Chapter 26

Tattooed Rebels

This was how the world worked
back then. Sound spoken in reverse
to mislead another enemy
treading close. I disguise myself
as an animal so we may escape. Home
the only place where I utter your name
safely. The story of our lives sewn
into tapestry so we never forget
who we are. Every syllable needlework
snagging into skin. How the scarf weighed
a world on our shoulders as we ran.

—Alyza Taguilaso, *Sayabovry*

ZAP!

Theos appeared out of nowhere.

"You're quick," Violet whispered, letting him into the house.

"I started the preparations after our talk," Theos said, putting down a white sack. "We must do this first before anything else."

She looked at the bag. "Let's do it in my room." She looked warily at the AI housemaid, who was just standing by, awaiting her command.

Theos nodded and followed her.

"Were you able to get it?" Violet asked as soon as they were alone in her room.

Theos looked around. "No cameras in here? Or AI?"

Violet frowned. "None. Your mother placed cameras in your room?"

"Oh, yeah." Theos shrugged. "That's why I just go out a lot."

"You know she can still track you, right?" Violet said.

"That's why I prefer to fly," Theos answered, sitting on Violet's couch. "Without a wristband." Then he reached down into the bag and produced a gold apple pin. "Here. Mom just officiated another moon ceremony. It's past midnight, Violet. She'll never know it's missing until the first light. We have to be finished by then."

Violet nodded. "The sooner the better."

"Ready?" Theos said.

Violet stood frozen in her own bedroom. She nodded. She had already agreed to this, and it was one of the most foolproof ways to regain their memory, just in case their parents put them under the Memoir Edit.

"What are you waiting for?" Theos asked as he took out the tattoo pen from the sack.

Violet gulped. Theos laughed at the sight of her, looking wary.

"It's clean. The needle's been sterilized," Theos assured her. "This pen, well, is Earth tech. Got it from a trader during one of my adventures."

Earth tech and traders. Now, those could be good stories for later. But Violet said nothing as she sat beside Theos. They must hurry. For all she knew, her parents could be dissecting Exa by now and erasing Mrs. July's memories. She bared her shoulder.

"Hmm," Theos said. "I don't think the shoulder is a good idea."

"Back?" Violet asked.

"Somewhere you can view easily without others seeing," Theos said. "What do you usually do first thing in the morning?"

"Shower?"

"Breasts?" Theos said.

"WHAT?" Violet gasped. "YOU MANIAC!"

Theos guffawed.

"My nape," Violet said. "I tie my hair up in the shower sometimes. I'll see it in the mirror."

Theos, still laughing, pushed Violet's hair out of the way. Violet positioned herself on the couch, bending her head.

"It's just a simple QR code," Theos said, now serious.

"How do you know how to tattoo a QR code?" Violet whispered.

"That's a secret." Theos smirked, his breath brushing on the hairs of Violet's nape.

WHIRRR!

Violet squeezed her eyes tight through the tactile pain. "Ouch," she silently groaned.

"Sorry, sorry," Theos said, but didn't slow down.

After a few minutes, Theos wiped the surface of his masterpiece: a simple QR code on Violet's nape.

"Try it out," he instructed Violet.

Violet raised her wristband to the level of her nape to scan the QR code. Suddenly, a holographic file appeared.

"There!" Theos beamed. "It works! You just need to open that file."

Violet nodded. "Where should I tattoo yours?" She took the tattoo pen from Theos.

Theos bit his lip. "My thigh?"

"MANIAC!" Violet rolled her eyes. "I'm not kneeling in front of you!"

Theos laughed. "I guess nape will do as well."

Violet rolled her eyes again as Theos positioned himself. She gently moved his long dark locks aside and positioned her untrained hand on his neck. She sighed nervously.

"Should I practice somewhere else first?" she muttered.

"We don't have much time," Theos reminded her. "Just copy the pattern."

Violet sighed, her breath slowly brushing against Theos' exposed nape. She brought the tip of the tattoo pen on Theos' skin.

"You can do it," Theos gently encouraged her. "It's easy. Wipe off the blood afterward."

Then she realized, as she was doing Theos's tattoo, that this was the first time anyone else had been in her room. Not even the twins had been here. She felt an unexpected tingle in her spine. She brushed it off.

Soon, Theos found himself trying to admire the tattoo on his nape in Violet's mirror, angling just a bit as it was hard to.

"If it ever comes down to that, promise me you'll remind me of the tattoos," Theos told her.

"I will," Violet said, covering the tattoo with the collar of her jacket. "Now, let's go."

ZAP!

The two teleported outside the Tree of Knowledge of Good and Evil.

Theos sighed as he took out his mother's gold pin. "Here goes nothing," he muttered under his breath.

Slowly, they climbed the steps to the entrance. Theos flashed the golden pin to the scanner.

The holographic doorway opened. The two looked at each other in relief.

"Where to?" Theos asked as soon as they got in.

The lobby of the Tree of Knowledge of Good and Evil was empty and dimly lit. There were no signs of AI robots in this area.

"We probably should get Exa first?" Violet suggested. "I'd rather we didn't split." She looked over at Theos, who looked so much in awe of the place. This guy could ruin stuff if left unattended.

"Where?"

"I have an idea," Violet said. "They will likely run tests and examine Exa's genetic sequence." She started walking.

Theos followed her. "Who do you think is here?"

"I believe my parents, some AI robots, mostly humanoid scientists, and no one else," Violet replied. "It's just after midnight. Why would anyone be here at this ungodly hour?"

"To capture fugitives," Theos whispered. "I have to admit Agham is pretty smart to do it now. My mother didn't see it coming."

Violet hushed him as they crossed several floors. Then, around the corner, she heard familiar voices. Theos suddenly grabbed her, and the two pressed their backs to the concave walls.

"… rerun it," Agham was saying.

"Agham, I already did!" Violet heard her mother say. "There's nothing. And it's okay—"

"It's not!" Agham angrily insisted. "We'll have to defend Project Minus."

"You know that there is a 0.001 percent chance that an infant may be excluded from Project Minus," Azula reminded him.

"And what are the chances that Exa July is part of that?"

"Looking at the genetic makeup," Azula said, "a pretty big chance she is part of the 0.001."

Theos looked at Violet and whispered, "I told you she's normal!"

Violet looked irritated. "I never said she wasn't!" she whispered back. Then she put her finger against her lips to shush him.

"Let me just copy the file," Azula was saying. "Then we can proceed with Nia July."

"Then what do you propose to do with the twins?" Agham said. He sounded stressed and desperate.

"Let the twins go," Azula said. "Let Case Number 3 go."

Violet's eyes widened. Was this the end? Exa and Alec were innocent!

"Give them back their Eden status," Azula added. "Your fight is with Alexander and Miriam Leons for not trusting the system. We can put it at that—"

"The Council will not have that," Agham angrily insisted. "A moment of weakness or doubt can ruin people's trust. And we've been hunting Cases 2 and 3 for so long."

Azula said something low that Violet could not hear.

"Regardless, Agham will have to inform the Council,"

Theos whispered.

Violet nodded.

Then they heard footsteps getting closer.

"Duck!" Theos whispered and pulled Violet down to the floor with him.

Agham and Azula exited the spherical chamber.

"I was hoping to do this with Violet tomorrow," Agham said.

Theos looked at Violet pointedly. Violet looked away. They were so close. Their faces were only inches from each other.

"You can do half now. Erase the essential ones," Azula said, her voice fading as they walked away. "Then have her do the half. It's not really punishment…"

Violet and Theos stood up when they were alone.

"Can you do it later?" Theos suddenly asked Violet.

Violet sighed as they stepped toward the spherical chamber. Exa lay peacefully on a silver capsule, wearing a thin white hospital gown.

"Do what?"

"Erase Exa's mom's memories," Theos said.

Violet sighed again. "I don't know," she answered truthfully. She went straight to the holographic screen beside Exa.

Theos went behind her. He took Exa's hand. "We will get you out of here, Exa," he whispered, then kissed her hand.

Violet raised an eyebrow but didn't say anything about it. "Wake her up. We have to teleport her."

She found the medical file her mother had just done on

Exa. She copied it to her wristband.

"Where to?"

"To where her brother is," Violet said simply. She looked glum. "But honestly, I don't know where he is."

Theos watched her closely and looked at the time. "You know what? You wake her up. She needs you more. I'm going to try and stop your parents."

Violet spun around to face him. "What? That's dangerous, Theos!"

"We already know this mission was dangerous to begin with," Theos answered. "Anyway, I'm just going to create a distraction. After Exa is safe, come and get me. I mean, you can't leave me alone."

That made Violet smirk. "What made you think so?"

Theos feigned hurt. "After what we've been through? We're tattoo buddies!"

Violet couldn't help it. She laughed.

Theos smiled at her and patted her back. "There, that's better," he said. "I'm going now." Then he went away.

After copying Exa's file, Violet turned to her friend and began shaking her. "Exa? Exa? Come on, we have to go."

She looked around, knowing that no one was with her, but she couldn't shake a certain feeling that she was being watched.

"Exa!" She shook her friend again. Nothing. She looked around. Then she saw something.

She quickly took a gas mask and put it on Exa's face, configuring it to mimic the smell of ammonia.

Within seconds, Exa started to cough and open her eyes.

"Exa!" Violet exclaimed in relief. She hugged her friend.

"Violet?" Exa gasped. "Violet!" She hugged her back. "Where are we?"

"The Tree of Knowledge of Good and Evil," Violet answered, and then she told her everything.

"We have to save Mom!" Exa exclaimed, hopping down from the silver capsule.

"No!" Violet stopped her. "You have to go! I can teleport you out! Oh, you have a wristband. How?" She frowned at her. "But you have to go to Alec. I have your medical files. You're okay. Aside from heterochromia, you have no other anomalies."

"But Mom—"

"Theos and I will handle her."

"Theos is here!" Exa's eyes widened a bit. A small smile tugged on her lips.

"Yes, we will handle your mother—"

"NO!" Exa loudly insisted. "I WILL SAVE MY MOM! What if she forgets about me and Alec?"

"WHAT IF I LOSE YOU AND ALEC AGAIN?" Violet exclaimed. "Do you know how many times I tried to save you? Planning and all that sneaking?"

"Your father is a scary man." Exa frowned.

"YOU THINK?"

Exa's face softened. She took Violet's hands. "You will always be our friend. We will always be your friends. Whatever happens. Fugitive or not. But this is my mom."

Violet sighed and nodded. "You're too stubborn. Fine." She wiped her eyes. "I would do the same if I were in your place."

Exa smiled. "Come on, let's go!"

Chapter 27

What Transcends Time Are Memories

I play the sandgrains back.
Two glass pyramids, summits joined
Measure moments that glissando down
The minute hole and even out
The heavy disquiet of the hour.
No single grain will root or flower:
Leaves, petals, birds, and rain,
Mother, friends, and lovers
Down the drain
Until—
I flick to reverse the course of things
And play the sandgrains back.

—Marjorie Evasco, *Hourglass Symphony*

Before first light.

That was Patia's instruction to Yvan.

"I think it's best to assume we weren't followed," Harold said. "There were no AI robots roaming around."

Yvan agreed. "We can take you to Heph to get those lasers off."

Harold smiled. "I'll meet you in front of the Tree of

Knowledge of Good and Evil." He turned to Liam.

Liam just nodded. Alexander, Miriam, and Alec would go inside to save Exa and Nia.

Alec only slept for a few hours, but now all he could think about was storming the Tree of Knowledge of Good and Evil and coming out with Exa and Nia.

ZAP!

Harold, Yvan, and Ysaac teleported back to Pangea.

Liam looked at the family with him: husband and wife holding hands, their estranged son looking away. He sighed. "Well, let's go."

"Here." Violet pulled Exa in the other way.

"Have you been here before?" Exa squinted at her.

Violet didn't meet her eyes. "My mom showed me." No need to tell Exa that she was expected to be part of the Tree of Life Council in the future.

As they neared another spherical room, they heard whirring and humming. Then, they felt a chill down their spine when they heard a familiar voice.

"He probably didn't think I would notice," a melodious voice said.

Exa and Violet looked at each other in horror. It was Levida Luna. Violet frowned, thinking of the golden pin Theos has stolen. How did Levida Luna get here?

Exa pulled Violet back.

"Thank you for reporting promptly to me, Levida." It was Agham. "My wife is already attending to Nia July."

Exa closed her eyes in horror as she bit down a frustrated groan. Violet squeezed her hand. Azula must have moved Nia July somewhere, but where?

Exa immediately activated her wristband and sent a message. Violet looked at her questioningly. She mouthed Alec's name. Violet nodded. She looked at her own wristband. It was five in the morning already.

"I'm surprised you are amenable to this," Agham said. "But rest assured, it is safe. I know you treasure your son as much as I treasure my daughter."

Violet gasped. Theos!

"Of course," Levida said. "Anything for Eden. But I'm telling you, my son never works alone—"

"I am telling you, Levida, no one was with him when we caught him trying to get to Nia. Azula was one step ahead, changing the location of the other Memoir Edit machine," Agham said. "Your son was alone."

Levida said in a quiet voice, "You might want to investigate your own blood, Agham."

Agham grunted. "I'm positive my Violet will also do anything for Eden."

Levida nodded and purred. "Of course, of course, her destiny is also written, like my son's."

Exa looked at Violet. She mouthed, "What does she mean?"

Violet tried to look away and shrug, but Exa did not buy it.

"If I am too willing to put my son under the Memoir Edit, I'm sure the head of the Council is, too, with his daughter," Levida said silkily.

Exa's eyes widened on Violet. "They want to erase your memories? Why?" she gasped. Then she realized. "You will also be part of the Council, too. In the future."

Violet didn't dare meet her friend's eyes. She didn't know how to explain to Exa how complicated her family can be. "I'm sorry," she muttered quietly. "About my father. He was the one who ordered the robots to attack you. To take your parents hostage."

"It's not your fault!" Exa whispered. "It's Theos! It's his mother! It's your father!"

"I wasn't able to do anything—"

Exa grabbed Violet's arms and forced her to look at her. "YOU. ARE. NOT. YOUR. FATHER."

Suddenly, Levida spoke. "I heard something, Agham." She sounded suspicious.

Violet and Exa ducked their heads. Exa covered her mouth.

"I'm sure you are only hearing things after getting so little sleep, Levida," Agham said. Then he offered, "How about let's drop in on Azula and Nia?"

"Of course," Levida said. "Theos will take a few more minutes, right?"

Violet and Exa tried to make themselves smaller in the corner as they heard footsteps coming closer and then moving away. When Agham and Levida were gone, they rushed into the spherical room.

The image in front of them confirmed Violet's suspicions.

Theos was strapped to a gold capsule and angled almost upright. He wore a golden metal helmet, and in front of him, holographic images in red were shown. In front of him was the

Memoir Edit, to which he was connected.

"She really tampered with her son's memories!" Exa gasped angrily. "Unbelievable!"

Violet started to hyperventilate. She knew this was a possibility, especially given the last emergency Council meeting Theos had told her about. But she always thought they could get away with it. Now, seeing Theos in this state, she knew she would be next.

She rushed to Theos's side. "THEOS! THEOS!" she hissed, shaking his arm and slapping his cheek.

But his eyes stared ahead, dilated. His lips were slightly agape. How come she only noticed now, for the first time, that he had light gray eyes? And they were striking.

Violet rushed to the Memoir Edit machine. She started pressing on the hologram, but nothing was happening.

"It's locked. The settings are locked," Exa said. "I already tried."

Violet sighed angrily as she looked at the defenseless Theos. There was no hope for it. She just had to remind him of their tattoos. His last words before he left were ringing in her head. *I'm sorry, Theos.*

Exa went to Theos's side. "We will come back for you, Theos." She squeezed his hand.

Violet turned to Exa. "There's no saving him now. Let's save your mom."

Exa followed Violet, and repeated her question. "Tell me the truth. Will they do that to you too?"

Violet nodded. "I don't know yet, so let's get this over with before they catch me."

Exa caught her arm. "Wait!"

Violet turned impatiently. "What?"

"How will you remember us?" Exa asked her, almost panicking.

Violet sighed, thinking of her tattoo. "I'll manage, I promise." She turned and almost sprinted away. Exa followed her on her heels.

"I will make you remember," Exa promised.

Violet smiled. "I know you will."

Just as they turned to the next corner, they heard gunshots.

The holographic entrance to the Tree of Knowledge of Good and Evil scanned Edmund Siegart's gold pin without issue.

Alexander, Miriam, and Alec were all carrying rifles. Alec made a mental note that a handgun might be much easier to carry in these kinds of missions.

His wristband made a tiny sound as soon as they stepped in. His eyes widened when he saw that it was a message from none other than his sister.

ARTEMIS:

4th floor, 3rd sphere.

"She messaged me. This is where we need to go," Alec said, showing the others the message.

"Patia said fifth floor and second sphere," Alexander said. "Her reminder is that they will transport them before first light.

We arrived earlier."

"Well, they're already here," Alec said. "How else could Artemis message me?"

Miriam and Alexander exchanged glances.

"Listen, it could be a trap," Alexander said. "They could have gotten her wristband and sent it to lure us."

Alec looked at him with an annoyed expression. "And what if this was really her?"

Miriam sighed. "We can check it out. It's on the way." Alexander agreed, and they started moving.

They found a holographic elevator. "Where are the people?" Alexander said.

Miriam and Alec looked around warily. But it was as if the Tree was empty.

"Probably because it's still early?" Miriam suggested.

But as they reached the fourth floor, they suddenly found themselves amid a group of AI robots.

"Intruders on the fourth floor, intruders on the fourth floor," the AI robots began chanting and pointing with their laser guns.

Surrounded, the three had no choice.

"Fire!" Miriam shouted as the three of them started shooting.

PING! PING! BANG BANG!

"Look out!" Alec shouted as a red laser beam passed beside him, missing his shoulder by an inch.

"No need to worry about us!" Alexander said as he and Miriam stood back-to-back, shooting the AI robots. There were at least thirty of them.

"This must be the right floor," Miriam said. "Why else are they all here?"

Alec tried to target the necks and joints, but too often, his bullets missed and clanged upon metal.

"Get the laser guns!" Alexander instructed.

Alec dived for a laser gun from a broken AI robot. He started shooting. His parents did the same.

Now, too many lasers were firing all at once.

"We have to move!" Miriam said. "Apollo, get over here with us! We will cover your back!"

Alec did as he told, moving along with Alexander and Miriam, slowly but surely, while trying to avoid the laser beams.

Suddenly, a familiar voice shouted, "MOVE!"

A spherical metal ball rolled onto the floor amid the AI robots. It whirred and glowed purple. In another five seconds, it emitted a bluish-purplish static lightning, electrocuting those within its reach.

Alec stared in awe. The ball had killed at least three robots at once.

They looked. It was Violet, and she was preparing to throw another ball. Beside her was Exa, who had already grabbed a laser gun.

"Violet!" Alec breathed. "Exa!"

Violet smiled at him. There were so many things she wanted to explain to him. To tell him that she was on their side.

"It's Artemis!" Exa rolled her eyes at him.

Violet looked confused for a moment.

"Code names," Exa muttered.

Violet threw another ball to the other side, killing another three robots.

"We're so glad you're okay!" Miriam beamed at Exa, who merely stared at her with a hard expression.

"We still have to save Mom!" Exa reminded them. She glanced at her twin and a look of understanding passed between them. So these two are their real parents. Exa sighed, trying not to think about it now. So much for a reunion.

"Of course, of course!" Miriam fussed a bit, her hand mid-way of reaching out to Exa, torn about hugging her daughter for the first time, and the ongoing battle.

Violet threw another ball. Thank God she had listened to her mother when she discussed these little toys, and thank God they were just up on the shelves that she and Exa were passing when they heard the shots.

"Come!" Violet instructed. "I think I have an idea where she is!"

Suddenly, more AI robots rounded in the corner, pointing their laser guns at them.

"They must have called for reinforcements!" Alec said angrily.

"Go!" Alexander said. "Go with them, Apollo! Your mother and I will hold them off!"

Alec nodded as he sprinted after Violet and Exa. The three friends ran down the corridor, past several spherical rooms. Looking at each other, they almost smiled.

They were together again. Just like old times

Chapter 28

The Attack on the Tree of Knowledge of Good and Evil

But if we leap up over the rapids, then our legacy will instead be all the suffering and joy of that single moment, the fear and exhilaration of putting everything at risk. All the actions of our lives, the choices we make, and the challenges we accept, will go to make the bones and flesh of our children, becoming the substance of those future lives. And that's why we mustn't take the easy way.

—Ahn Do-hyun,
The Salmon Who Dared to Leap Higher

Running side by side, the three friends were elated for the first time since the twins' eighteenth birthday. Of course, as expected, it was short-lived.

Exa heard it first: the silky voice of Levida Luna just a few steps ahead of them.

"…and to do this on twilight, why, I guess it seems a good idea…" the councilor was saying.

She immediately skidded to a stop and instinctively pulled Violet back.

"Hey!" Violet groaned.

"Did you hear that?" they heard Levida ask.

"Shhhh!" Exa said as she pushed Violet into a random spherical room. Even Alec looked bewildered as he followed them. "Levida Luna can't know you're here!" Exa hissed.

Violet suddenly understood. If she were caught, her father would have no choice but to put her under Memoir Edit like Theos.

"Why?" a clueless Alec asked.

"Stay here!" Exa instructed Violet. "Alec and I will deal with them. Once the coast is clear, go to Mom! Save her! We will take her to Pangea!"

"Pangea?" Violet repeated. The name seemed familiar to her. Oh, right, it was the name of the unified piece of land consisting of all of Earth's continents. A historical geographical name.

"It's an underground world." Alec winked and smiled. "Maybe another time, we can take you there."

Violet shook her head, bewildered. "What? An underground world?"

"Miriam and Alexander are probably on their way too. We have help," Alec said, getting back to the matter at hand.

"Our biological parents," Exa quickly explained to Violet. Then she became distracted, squinting at the machine behind Violet. "Oh, this is the machine I saw in Isle Luna."

Violet and Alec looked. Yes, Violet thought, remembering her mother's lecture. That is indeed that machine. "It temporarily modifies a person's genes, using the genes of other organisms, to give them temporary ability, like Theos's flying. Levida must've been using this machine regularly on him to sustain his flying skills. I haven't had the chance to show him."

"What happened to Theos?" Alec asked.

The girls ignored him. "Not just bird's flight," Violet said, "but other abilities you can think of. Like an eel's electric shock."

Suddenly, a loud alarm cut through the air, a shrill and terrifying wailing. Then, Azula's voice was heard. "THE TREE OF KNOWLEDGE OF GOOD AND EVIL IS UNDER ATTACK. ALL HUMANOID SCIENTISTS DEACTIVATE AND LOCK. ALL AI MILITARY ROBOTS DEFEND THE TREE AND CAPTURE THE INTRUDERS."

"They know!" Violet gasped. "Of course they know!"

"Stick to the plan!" Exa said quickly. "We'll come find you! Just don't let Levida know you're here!"

"And save Mom!" Alec added.

"How are you going to escape?" Violet exclaimed. "Have you thought this through?"

"We have people waiting outside," Alec explained.

"People?" Violet asked, confused.

But the twins were already gone, running up ahead and shouting, deliberately making noise.

"I REPEAT," said Azula's voice over the PA system, still loud and clear. "THE TREE OF KNOWLEDGE OF GOOD AND EVIL IS UNDER ATTACK…"

Violet heard Levida Luna's and her father's voices.

"…MENACES! FUGITIVES! HOW DARE YOU ATTACK US!" Levida was shouting.

"CALL FOR REINFORCEMENTS!" she could hear her father shout.

She couldn't leave this chamber. Not yet. She couldn't be

seen by Levida and her father.

Violet looked at the machine. Suddenly, she had an idea. She turned it on. A hologram monitor came to life. She looked at the settings. She smiled and started it. Then, she positioned herself in the center. Pink rays suddenly emanated from the machine, showering her in its glow. Violet felt it on her skin. It was hot. Uncomfortable, but not painful. She felt it creeping, like it was forcing her pores open.

Then she heard many footsteps. Another surge of AI robots ran past her, not heeding her inside the chamber. She had to help her friends. She had to free Mrs. July.

Ping! Ping! Ping!

"Catch us, over here!" she heard Alec shout.

Then Levida Luna shouted, "I have to get my son! Theos! Agham, this is unacceptable! The Council will hear of this!"

"YOU DON'T TELL ME WHAT IS UNACCEPTABLE! IT WAS YOUR SON WHOM WE CAUGHT HERE!" her father thundered.

Violet heard them running away in the other direction.

The machine wasn't done with her. But she had to act now. Violet stepped out of the center and turned off the machine. She could feel it in her fingertips. Electricity.

She held out her hand. To her amazement, purple electricity visibly crackled at her fingertips. She smiled. Then she ran from the chamber.

She had to get to Mrs. July.

"AHHHHHHH!" She heard Exa's scream. She stopped in her tracks. *Exa!*

The number of AI robots was overwhelming. Azula Malaya's

voice continued to echo all over the Tree.

"...DEACTIVATE AND LOCK. ALL AI MILITARY ROBOTS DEFEND THE TREE..."

Two AI robots had disarmed Alec, shooting away his gun before effortlessly grabbing each of his arms and lifting him into the air.

Levida Luna had run in the other direction, back to where they came from, running for the son whose memories she had modified. She looked deranged as she picked up her white garments and ran away.

Agham furiously spotted the twins. He marched toward them, roaring, "YOU!" He suddenly grabbed Exa by the hair.

"AHHHHHHH!" Exa screamed.

"EXA!" Alec screamed. There was no use for code names now.

"HOW DID YOU GET AWAY FROM YOUR CAPSULE?" Agham thundered. "WHO LET YOU OUT!"

Exa struggled, trying to kick Agham. But with formidable strength, Agham lifted her from the ground by her hair.

"AHHHHHHH!"

"NOOOO!" Alec kicked the robots' knees, instantly weakening them. They loosened their hold on him, and Alec jumped, pushing Agham.

"ARRRGHHH!" Agham stumbled backward, releasing Exa.

Several AI robots ran toward them, pointing their laser guns.

"GET A GUN!" Alec shouted at his twin.

They managed to get a laser gun each, but they were outnumbered.

Then, they heard a familiar voice. "No!"

But before they could look toward the source, suddenly there were laces of electricity running and spreading across the floor.

"LOOK OUT!" Alec shouted. He and Exa staggered backward to escape the purple electricity.

Agham's eyes widened in shock, and immediately, he stood up so that only his rubber boots were in contact with the ground.

Instantly, all AI robots in contact trembled violently as purple electricity climbed over them, mercilessly sending them into a short circuit. Then, almost in unison, they collapsed.

Agham shouted in frustration. It was the loudest sound the twins had ever heard. They looked toward the source of electricity, but there was no one there. They looked for a metallic ball, like the one Violet had thrown before, but there was nothing.

Alec looked at Exa knowingly. That voice belonged to Violet. But where was she? Hopefully, she was now saving their mother.

Agham turned to the twins. "I'VE HAD ENOUGH OF THE TWO OF YOU! YOU TWO WERE SUPPOSED TO BE DEAD!"

Exa pointed a laser gun at him. But Alec cautioned her. "We can't hurt him. Think of Violet. We've won anyway."

Unhinged, Agham shouted again as the realization hit him. He bent over, detached a big piece from an AI robot's body, and threw it over toward Alec and Exa.

"LOOK OUT!" Exa shouted.

"LET'S GO!" Alec said. "RUN!"

Agham continued to shout. But the twins had started running away from him, leaving him in his blind rage.

———✦———

"Let's go!" Exa said, running with Alec.

Suddenly, the voice in the PA system changed. It was Agham Malaya. He sounded stern and calm. Like the Head of the Tree of Life Council that he was. "ALL AI MILITARY ROBOTS FIND AND ARREST THE INTRUDERS." His voice was contrary to the latest state the twins had encountered him in.

The twins continued running the halls of the Tree of Knowledge of Good and Evil. Out of their periphery, they spotted hordes of AI robots marching, looking for them. Some were running after them.

Ping! Ping!

The twins tried to avoid the lasers. They ran sideways or ducked, until they reached a dead-end chamber.

"Is this it?" Exa exclaimed.

"Maybe," Alec said. "There's no other way! Turn right!"

"Where is Violet?" Exa exclaimed.

As they turned right, a locked chamber greeted them. Alec shot at the lock with his laser gun.

Ping!

A holographic entrance opened.

"MOM!" Exa shouted.

Inside, Nia July was strapped to a gold capsule, similar to Theos's. The gold capsule held her body almost upright. A familiar metallic cap was on her head. Her hazel eyes were dilated and staring ahead.

"NO!" Exa screamed as the twins ran to their mother. They started pulling off the straps. "MOM!"

The only difference from Theos's situation was that there were no holographic images playing in front of her now.

Alec carried his mother's weight as they tugged at the straps. Outside the chamber, the sounds of footsteps were getting closer. But these footsteps weren't rhythmic. They were chaotic.

Then gunshots. *Ping! Ping!*

"Exa!" Alec warned, gesturing toward the door.

Exa grabbed again her laser gun, but she went for the little computer connected to the metallic cap. It was already blank. She felt tears stinging her eyes. Was the Memoir Edit done? Were they too late?

The sounds of the footsteps grew nearer. Exa pointed her laser gun at the entrance, ready to make their final stand.

To her relief, Miriam and Alexander came running in.

"We finally found you!" Miriam gasped in undeniable relief, running even faster to hug Exa, who instantly pulled away.

Alexander was almost breathless. "Patia's instructions were right!" He looked at Alec. "This is the second sphere on the fifth floor!"

Something registered in Alec's mind, but he stayed silent. His eyebrows even knitted in confusion as he pondered Alexander's words.

"Come on!" Miriam said, pulling Exa and gesturing to Alexander to help Alec and Nia. "Patia said there was supposed to be an exit right this way."

The twins followed Miriam and Alexander as they made

their way deeper into the chamber.

"You don't know how many AI robots we have had to kill to get to you!" Miriam complained. "We nearly ran into Levida Luna and her son!"

Then, to their surprise, Miriam opened a seemingly normal cabinet door. Inside was a chute.

"Liam and Harold should be waiting for us at the end. Then we teleport back," Alexander said.

Agham Malaya's instructions continued to play through the PA system. The sound of another set of footsteps were getting nearer. They were rhythmic this time.

The twins looked at each other and nodded.

"Don't shout," Miriam sternly reminded them.

Quickly, and one by one, they all jumped down the chute. Nia was held by Alexander and Alec. Their bodies were almost distorted as they fell on top of each other. But no one dared make a sound.

It was almost a free fall. Exa closed her eyes as she felt Nia's weight on her. She felt her insides curling up. Then wham!

"Ooof!" she couldn't help groaning.

They finally landed on soft sand. They were outside the Tree of Knowledge of Good and Evil.

All groaned, except for Nia July, who looked entranced, unblinking.

Liam and Harold rushed to them.

"Nia!" Harold shouted, hugging his wife. "Wait, Nia? Nia? Nia!"

"You took a long time!" Liam scolded them. The sun had risen. The sky was bright pink again. "Other scientists have started arriving for work! But thanks to Agham, they teleported away upon hearing that the Tree is under attack!"

Miriam started pushing them off the ground quickly. She looked up the chute. "We can chat later! Hurry! Let's go!"

Liam started helping them to their feet. "Safe house first, then Pangea tonight," he declared.

"Nia! Nia!" Harold kept shaking his wife, oblivious to his surroundings. But Nia, although breathing normally, was still somewhat nowhere with them. The twins looked on in horror.

Exa started crying.

"We didn't see Violet there," Alec whispered.

Exa blinked. "What?" she whispered almost incredulously. Had Violet intentionally left them? Why didn't she save their mom? Had she saved them from the robots and her father?

Then Alec, with narrowed eyes, looked at Liam, Miriam, and Alexander. "Funny, we also never saw or met any Patia there."

Alexander and Miriam realized that this was true and exchanged a glance.

Liam waved his hand. "We can talk about that in Pangea. Let's go. NOW!"

In another second, three zaps were heard. Alec brought Nia with him. Exa brought with her Harold, and Liam brought Alexander and Miriam.

The desert fell silent. The wind blew the sands to cover their footsteps. As if nothing unusual had happened.

Chapter 29

Back in Pangea

He catches his breath at the shimmer of wings when the birds shake droplets loose from their feathers. A light breeze passes through the Nipa fronds on the riverbank; fetches faint sounds of a church bell calling the faithful to prayer

It is time to come home.

—Marjorie Evasco, *It Is Time to Come Home*

It'd only been two hours since they reached the safe house. Everyone was silent and crestfallen.

The twins looked worried, side by side. Harold was kneeling in front of Nia, whom they laid down on the couch. She wasn't blinking at all, and Harold closed her eyelids in the meantime.

He kept muttering, "Please, please."

Alexander returned from the outside. He had just patrolled the perimeter of the safe house. "No robots," he informed them. "Liam, surely the elders will understand." He gestured to Nia and Harold. "We have to go."

Liam glumly nodded. "We need something like ammonia. Maybe that can wake her up. But all my herbs are in Pangea."

Exa found herself nodding, thinking about the smell that

Violet had given her to wake her up. The thought brought up Violet again. She groaned, not knowing what to make of her friend.

Zap! Zap! Zap!

Within a second, they had reached Casiah Desert and opened the glass cube to the entrance of Pangea. Then they made their way down.

Alexander and Harold carried Nia. The rest somberly followed them.

As soon as the familiar earth of Pangea greeted them, Liam began shouting, "Medic! Medic!"

William, the lanky doctor from before, heard them and came over. Liam explained to him quickly what had happened.

Several people rushed to them.

"Liam! Miriam! Harold! Alexander!" They all gushed, crowded, and hugged them home.

Many people rushed to help them care for Nia. Some brought a cot for her.

The elders started to meet them one by one. Ajax, Nora, and Yvan were almost running.

"What happened?" Ajax demanded, seeing Nia's state.

Liam filled him in. William came back with several medical herbs and potions.

Liam took a bottle and uncapped it. Several people recoiled from its smell. He dangled it under Nia's nose.

Soon, Nia started coughing.

"Nia! Nia!" Harold shouted.

Then she blinked. Harold hugged her suddenly. "Oh, I

thought I lost you! Nia!"

There were murmurs of relief and gratitude from the crowd. Alec and Exa smiled at each other and hugged. Then they stepped toward their foster parents with outstretched arms.

Then suddenly, Nia laughed. "Oh, Harold, what has gotten into you?" Then she looked around. She found the twins. She stared at them. "Oh, hi, uhm, who are you?" She smiled.

Alec and Exa froze.

The crowd gathered around them grew silent as well. Their looks of relief were replaced by disbelief.

Nia looked around some more and stood up. "Harold, where are we?" She tried to whisper.

"Nia?" Harold whispered. "What do you mean?"

Nia looked back at the twins. "Why are you looking at me like that? I'm Nia July! Nice to meet you!" She gestured at Harold. "This is my fiancé, Harold."

Exa couldn't help but release a painful moan. She looked away and sighed. Alec squeezed her hand. It couldn't be helped. The Memoir Edit had taken their mother's memories of them.

Chapter 30

A New Movement

Hypatia's case then was this. She lived in a time when her intellectual heritage, a seven-hundred-year-old tradition, was crumbling. The supports that had once seemed so secure—the Museum and the libraries—had all been swept away by the swell of ignorant dogmatism. Almost alone, virtually the last academic, she stood for the intellectual values, for rigorous mathematics, ascetic Neoplatonism, the crucial role of the mind, and the voice of temperance and moderation in civic life.

—Michael A.B. Deakin,
Hypatia of Alexandria: Mathematician and Martyr

As soon as Violet had released the purple electricity, killing the AI robots in one blow, someone suddenly pulled her away from the view of her father and her friends. A familiar voice whispered in her ear.

"I told you to play your cards right, Violet."

"Mom!" Violet turned around to face her.

"I see you have used the GeneMod," Azula noted, her eyes twinkling. She didn't look angry.

So that's the tech's name, Violet realized.

"I have to save Mrs. July!" she suddenly realized, trying to scramble away, but Azula caught hold of both her arms.

"No." Her mother's voice was stern and hard, but she was smiling. "There's nothing you can do."

They heard Agham shout in rage, then a clash! He shouted again.

"It's done." Azula smiled. "I've already taken her memories."

Violet stared in horror at her own mother. Oh no. Why? What will Exa and Alec say?

Azula saw the panic in her expression. She cupped her daughter's face. "Come," she said. "Let's get your father. It's almost over."

Almost over? What?

They heard footsteps. Violet peeked and saw the twins running away in the other direction, far from her father.

"They'll look for Nia," Azula said. "Their biological parents will meet them. Don't worry about them, Violet."

Violet looked at her mother. "What do you mean?" she whispered. "How do you know these things?"

Azula just smiled.

"The AI robots are everywhere!" Violet wanted to stomp her feet. "Call them off!"

Azula looked at her as if Violet had lost her mind. "And what would your father say?"

Their father shouted in anger.

"Come," Azula sighed, then pulled Violet with her, sprinting toward Agham.

"Agham!" she called.

Agham turned to them. He looked angry. He looked

ferocious. Violet shuddered behind her mother.

But Azula ran on and hugged her husband. "Husband," she breathed. "It is done. I have taken Nia's memories, but then Miriam and Alexander started shooting—"

Violet looked at her mother in confusion.

"They what!" Agham thundered.

"I had no choice!" Azula insisted. "Then I saw your daughter here! I don't even know how Violet got in here! I had to save Violet from these fugitives—"

"THEY WILL PAY FOR THIS!" Agham thundered. He turned to Violet. "AND YOU, YOUNG LADY, WE WILL TALK!"

Violet shuddered. She felt like a little girl again, berated.

Agham turned to his wife. "You are sure Nia has no memory of the twins? How about Exa's medical files?"

"I have them!" Azula said. "Nia's memories and Exa's files."

"Good," Agham said. "We have leverage." He turned to his wristband and connected with the PA system. "ALL AI MILITARY ROBOTS FIND AND ARREST THE INTRUDERS!"

"Let's go. Let's keep the family safe," Azula insisted. "Let's proceed to my office. We'll be safe there."

They started walking toward the head scientist's office. Violet felt uneasy that she wasn't running, unlike a while ago. The AI robots just ran past them instead of chasing her.

The head scientist's office was on the top floor, the only spherical chamber there. To Violet's surprise, another model of the Memoir Edit stood in front of her mother's desk. Her breathing hitched.

Agham's wristband alarmed. He sighed, turning to Azula

and Violet. "The Council is in unrest, as expected. I will call an emergency meeting." He turned to his wife. "Handle Violet. I will be with you two shortly, and then we will proceed."

"Handle…me?" Violet croaked. But her father left the room, and Azula locked it using her fingerprints.

She smiled at her daughter.

"Mom?"

"Sit down, Violet, and calm down," Azula said, smiling sweetly. "Silly, I will never let anything happen to you."

"What do you mean?"

"That's why we're doing this," Azula said.

Violet sat on the couch, confused.

"First of all, I am very proud of what you have become. You are brave, strong, and intelligent. You know what is right or wrong," Azula began. "But you have to be more careful."

"Exa and Alec are innocent!" Violet said. "You saw the medical file—"

Azula raised a hand to silence her daughter. "We know that. Listen, enough about the twins."

"This is about them—"

"No, it has always been bigger than them. Greater than a case of heterochromia," Azula said. "This is Eden."

Violet made an annoyed sound.

"And Pangea."

Suddenly, Violet's head snapped to her mother. "What? How did you…?"

Azula sat composed, her face stern. "Listen carefully, Violet. These are the consequences of your actions. Thanks to your relationship with the twins, the Council is in uproar. Everyone in the Council will begin to doubt your father. He needs us now. He needs our support.

"Don't you realize that Levida Luna is already on to you? It's good that you sought the help of Theos Luna, but I hope he remembers you after Levida put him under the Memoir Edit.

"Levida will seek to oust your father as the head. But your father, for your sake, for our family's sake, will turn the tables on Levida. After all, Theos stole from Levida, and both Levida and Agham saw him in the Tree. No one but I and your father saw you. I can already predict the meeting." Azula chuckled wryly.

"The Council needs to change!" Violet insisted. "There are so many unfair, unjustified laws!"

"Unfair, yes," Azula said. "Unjustified? No. Everything has its reasons. You still haven't trained in the Tree of Life Council, Violet. It's high time you do. Then you will understand.

"Don't you see, my daughter? This is our legacy! This is your legacy! You will become part of the Council soon! And we are hoping you'll replace your father! You will be his successor!" Azula's voice was rising.

Violet cringed.

Azula saw her daughter's fear and hesitation. "This is power, Violet. You are born into this. I've seen your strengths, my daughter. You can wield it even better than your father. I have told you that you know what is right and wrong. Just play your cards right."

Violet's face changed as understanding came to her. She looked less uncertain now, and she nodded at her mother.

Azula smiled. "Very good, Violet. It will take time. But don't worry. Right now, everyone needs time. Including your friends."

She continued, "I will delete all footage in the Tree of the Knowledge of Good and Evil involving you. That includes you taking Exa from my medical examining chamber, stealing electric balls, using the GeneMod, and electrocuting an entire floor of AI robots. No one can point a finger at you. I will make you a clean slate."

Violet could see where this was going. It made sense. After all, she knew this was almost inevitable. But she finally had a tiny sliver of illumination from her mother.

"The Memoir Edit is powerful tech. Do you know that not only does it take away, modify, or edit memories, you can also gain another person's memories?" Azula said.

"It can?" Violet asked. "But how would that help us?"

Azula smiled as she reached into her pocket, holding a tiny chip. "Password: Hypatia. It won't make sense. But for now, try to remember."

Then, the entrance opened. Azula quickly hid the chip in her pocket. Agham strode in, quite unhinged again.

"Those fools!" he stormed. "The Council is turning against us! Do they have any idea how great our efforts were?"

He continued to stride back and forth. "No, no, no!" He looked as if he were talking to himself and shaking his head. Then he snapped at Violet. "You will be part of the Council!" He looked at Azula. "Have you talked to her?"

"I have," Azula said, smiling. "She understands things now. She knows what needs to be done."

"Do you, Violet?" Agham turned to his daughter.

Violet swallowed. Her mother was right. Violet knew what needed to be done. "Yes," she said, her voice cracking.

Agham sighed in relief. "Thank you, Violet. You and your mother will help me. We will show these fools. Our family will prevail."

He offered his hand to Violet. Violet took it. She was surprised she wasn't shaking.

Agham pulled her up to her seat and gestured to the Memoir Edit.

Of course, Violet thought. It was the only way. She squared her shoulders as she stepped forward. "I'll go," she declared. "Whatever Eden needs."

Her mother smiled at her proudly. Her father looked stern and pleased at the same time.

Violet went into the golden capsule. Her mother walked forward to help her strap in.

To her surprise, her mother whispered, "I like your new tattoo, by the way." Then Azula carressed Violet's nape before she kissed her cheek. Violet tried her best not to shudder at her mother's touch. She looked forward.

Her mother put the metallic cap on her head. Within seconds, Violet's dark eyes dilated.

Glossary

AI. Artificial intelligence.

altimeter. A form of technology found in wristbands where the user can adjust the point in altitude where they can teleport to.

blue-tipped dolphins. Dolphins that have mutated to mimic a human's voice.

Casiah Desert. One of the dryest and hottest areas in Eden 2.0. It is not located on a floating island but is on the surface of the planet. It contains the entrance to Pangea.

Case Number 3. The unsolved case of a twin presenting with complete heterochromia who didn't make it to the intensive care nursery.

Case Number 4. The hypothesis that the heterochromia twins from Case Number 3 have infiltrated the system.

GeneMod. One of the pieces of technology developed by the Tree of Knowledge of Good and Evil which takes the genes of other animals and modifies them to give its users new abilities, such as electric touch and flying.

Ficus trees. Found prominently on Isle Luna, the Ficus trees, or *balete* trees, have mutated to have memory touch, to provide space inside them (that some residents use as their homes), and to facilitate easier transport or slower teleportation in the island through its vines, trunks, and roots, with the help of AI, according to other Eden scientists.

heterochromia. A condition wherein a person's irises are of different colors. Complete heterochromia is when the other iris has an entirely different color than the other. Partial heterochromia is when a part of the iris has a different color than the other parts of the iris.

Identifier. Another form of technology in Eden 2.0 which can identify different plants and animals in a given landscape.

intensive care nursery. Also known as Intensive Neonatal (Anomaly) Care. An intensive care unit where newborns are taken after they are assessed to be unhealthy, with anomalies, or with mutations. Rumors have spread that these newborns were killed and discarded instead of treated.

Isle Akalasa. Another floating island in Eden 2.0 which is characterized by its cold and snowy landscape.

Isle Luna. Known as the island of the moon because its inhabitants practice paganism or moon worship, it is one of the highest floating islands in Eden 2.0. It is also the territory of Levida Luna and her son, Theos Luna.

Isle of Templos. Another floating island in Eden 2.0 that houses abandoned temple ruins, which prohibits anyone from entering. However, since the island is directly above the Sophronia Ocean and near the *Voguer W*, it has served as another party spot for teenagers. Its abandoned temple has a water source inside it, which flows into a long narrow river that eventually turns into a waterfall.

the Memoir Edit. One of the newest pieces of technology developed by the Tree of Knowledge of Good and Evil. It can erase, edit, modify, or copy the user's memories. Some of its purposes are to free the user of bad memories that cause psychological trauma, and to modify the memories of prisoners to facilitate their corrective rehabilitation and expeditite their reintegration into the society of Eden.

mimosa. Also known as *makahiya*, mimosa flowers have mutated to open and close their petals continuously. In increased doses, its extract is being studied for the treatment of memory loss.

Pangea. An underground world, located underneath Eden. It lacks the AI technology of Eden and is reminiscent of the society and technology of Earth, the old planet.

Project Minus. A controversial project of the Tree of Life Council to address the overpopulation concern of Eden and to ensure that only superior species of humans survive, which they accomplish by exterminating newborns found to have mutations or anomalies.

rhinestone fish. A fish species found in the deeper areas of the Sophronia Ocean whose scales turn to rhinestone as it ages. Due to the weight of the stones, the fish swims very slowly. If one plucks a rhinestone, the fish's skin will emit a deadly venom that can seep through pores of the skin. The fish have to procreate before their whole body is covered in rhinestone and the fish eventually dies of hunger.

saltwater violets. Purple flowers which have mutated to survive saltwater depths and blooms only during sunsets.

Sophronia Ocean. The largest ocean in Eden 2.0.

the Tree of Life Council. Main ruling body of Eden 2.0, composed of high-ranking people whose ancestors have developed Eden into the paradise that it is. Besides ancestry, its members are chosen on the basis of merit, such as intellectual ability.

the Tree of Knowledge of Good and Evil. The main

scientific and technological body and building of Eden 2.0. They develop the latest technology of Eden 2.0, which comprises all areas of living, such as health, security, genetics, entertainment, and much more. The Tree is located adjacent to the Tree of Life. It is made of glass spheres situated inside another gigantic Tree that bears gold apples. Teleportation is forbidden inside as security measure. The entrance to the Tree needs a golden apple pin, belonging to the Tree of Life Council, to be scanned.

the Tree of Life. The blue glass building where the Tree of Life Council holds its meetings. Like the Tree of Knowledge of Good and Evil, it is situated inside another gigantic Tree. It is considered the most important building in all of Eden 2.0. It is located in the center of Eden, in one of the few islands on Eden's crust. Beside the Tree of Life is the Tree of Knowledge of Good and Evil. Teleportation is forbidden inside as security measure. The entrance to the Tree needs a golden apple pin, belonging to the Tree of Life Council, to be scanned.

the *Voguer W.* A legendary party ship docked in the middle of the Sophronia Ocean. It is not used for transportation but merely for partying by teenagers and their friends celebrating eighteenth birthdays, or simply as a spot to hang out.

wristband. Merely known as the wristband, it is given by the ruling body of Eden 2.0 to every citizen in order to track its population, to grant the ability of teleportation, to enable instant messaging and holographic education, and to place AI technology within reach.

Acknowledgments

Eden 2.0 began as a world built in my mind during my college years, with its pink moon and floating islands silently imagined, unwritten. For a long period of time, its idea lay untouched, unedited in my mind, as I moved on to being a doctor. Coming back to this creation has been illuminating, and therefore, let me acknowledge those who made it possible for me to even write this.

My heartfelt gratitude goes to my editor, Esther Abigail Daang. You were the first to step into this otherworldly creation of mine, and your skill and insight added magic to it. Thank you for your unwavering support since *Before High School Ends* and for instilling in me the boost and confidence to share Eden 2.0. We, authors, can only do so much, and it's editors like you who make the stories we write so richer!

I am deeply thankful to the following Filipino poets, whom I have met in my literary journey as a doctor-writer, and whose verses I have chosen as epigraphs:

Marjorie Evasco, "Hourglass Symphony" and "It Is Time To Come Home", *It Is Time to Come Home*, Milflores Publishing

Alyza Taguilaso, "Pearl" and "Sayabovry", *Juggernaut,* University of Santo Tomas Publishing House

Ralph Fonte, "Gunaw" and "Kartograpiya", *Ang Kartograpiya ng Pagguho,* Ateneo de Manila University Press

I am forever indebted to Professor Marj. She is my mentor from the Buenvenido N. Santos Creative Nonfiction Writing Fellowship and one of the first who encouraged my literary aspirations, despite being in a different field. She assured me that I can write a book, a sentiment I treasure until now. I have always loved to write, and bitterly thought literature and medicine are star-crossed lovers, until she, Dr. Joti Tabula and Dr. Lance Catedral made me see quite the opposite. They have given me guidance; hence, the push to go where my writer heart wants to wander and get lost. Of course, I am nothing without the encouragement of my other fellow writing buddies, most especially Dr. Ella Masamayor.

I have only met Doc Alyza last year, and not surprisingly, we both have common friends as well. Despite that brief time, I felt somehow closer to her as I believe I can see her in the beautiful, colorful, and vivid fragments of her poetry. Her words somehow leave me breathless and scrambling for more in the pages, and then quietly blending into insightful sentiments.

I first heard of Doc Ralph's poetry one afternoon where all the alumni of BNS Creative Nonfiction Workshop gathered together. His spellbinding words can transform into worlds, and I was captivated.

Their poetries stirred and resonated emotions in me, that somehow echoed a bit, or reminiscent of feelings when I was making Eden 2.0.

For turning my story alive in pictures, I have Cyrill Bagos to thank for. I believe your art can make Eden 2.0 feel closer to the minds and hearts of the readers.

Thank you to 8letters Publishing, Ms. Cindy Wong, Ms. J.K. Ojano, and the rest of the team for giving Eden 2.0 a platform. I am also grateful to D.A. Chan, The K Brothers, Patrick Flores, Joline Lim and the other writers of 8letters I have met during the MIBF 2024, whose advices I can count on, and whose words of encouragement can always lift me up.

About the Author

Mitchie Gonzales is a board-certified internist and allergist practicing in Lipa City, Batangas. When she is not looking after her patients, she writes poems, short stories, and creative non-fiction, aspiring to bridge medicine and literature. She is also the author of *Before High School Ends*, and her literary footprint include "Kuwento Bawat Kuwarto," in *Dx Machina 3: Philippine Literature in the Time of COVID-19* (2021), "If I Were God," in *Rx Narratives: Anthology of Creative Nonfiction of Filipino Internists*, and "Poetry in the New Democracy," in *Santelmo 7* magazine. She was a fellow in the third La Salle National Creative Nonfiction Fellowship for Doctors. She was a finalist of the 1st and 3rd Dr. Arturo B. Rotor Memorial Awards for Literature for creative non-fiction and short story categories, respectively.

Scan the code to explore more about the author's journey and upcoming projects.

www.ingramcontent.com/pod-product-compliance
Lightning Source LLC
Chambersburg PA
CBHW031344230926
51848CB00002B/373
9786211021447